Ergebnisse der Mathematik und ihrer Grenzgebiete

Band 58

Silvio Greco · Paolo Salmon

Topics in $\mathfrak{m}$-adic Topologies

Springer-Verlag Berlin Heidelberg New York 1971

Silvio Greco
Università di Genova Istituto di Matematica

Paolo Salmon
Università di Genova Istituto di Matematica

AMS Subject Classifications (1970)

Primary	13–02, 13B20, 13B99, 13C15, 13D05, 13D15, 13E05, 13F15 13G05, 13H05, 13H10, 13J05, 13J10, 13J99
Secondary	14A05

ISBN 3-540-05091-4 Springer-Verlag Berlin Heidelberg New York
ISBN 0-387-05091-4 Springer-Verlag New York Heidelberg Berlin

Printing: Zechnersche Buchdruckerei, Speyer · Binding: Konrad Triltsch, Graphischer Betrieb, Würzburg

Preface

The $\mathfrak{m}$-adic topologies and, in particular the notions of $\mathfrak{m}$-complete ring and $\mathfrak{m}$-completion $\hat{A}$ of a commutative ring A, occur frequently in commutative algebra and are also a useful tool in algebraic geometry.

The aim of this work is to collect together some criteria concerning the ascent (from A to $\hat{A}$) and the descent (from $\hat{A}$ to A) of several properties of commutative rings such as, for example: integrity, regularity, factoriality, normality, etc. More precisely, we want to show that many of the above criteria, although not trivial at all, are elementary consequences of some fundamental notions of commutative algebra and local algebra. Sometimes we are able to get only partial results, which probably can be improved by further deeper investigations.

No new result has been included in this work. Its only originality is the choice of material and the mode of presentation.

The comprehension of the most important statements included in this book needs only a very elementary background in algebra, ideal theory and general topology. In order to emphasize the elementary character of our treatment, we have recalled several well known definitions and, sometimes, even the proofs of the first properties which follow directly from them. On the other hand, we did not insert in this work some important results, such as the Cohen structure theorem on complete noetherian local rings, as we did not want to get away too much from the spirit of the book.

When we did not have the possibility of giving all proofs (and this happened, especially, in the central part and at the end of the work) we always tried, with very rare exceptions, to give precise references to the current literature. The reader will find it useful to consult the books which have already appeared on the subject, such as "Algèbre Commutative" by N. Bourbaki,

"Introduction to Commutative Algebra" by M. F. Atiyah-I. G. Macdonald, "Commutative Algebra" by O. Zariski-P. Samuel, "Eléments de Géométrie Algébrique" by A. Grothendieck and J. Dieudonné, "Algèbre locale. Multiplicitées" by J. P. Serre, and "Commutative Rings" by I. Kaplansky.

Content

Foreword

All rings we consider are commutative with an identity element 1. If $\phi: A \to B$ is a ring homomorphism, we assume that the image by ϕ of the identity of A is the identity of B; if A is a subring of B, we suppose that the identity of A is also the identity of B. If M is an A-module, we assume that $1m = m$ for any $m \in M$.

§ 1
Compatibilities of algebraic and topological structures on a set. The $\mathfrak{m}$-adic topology. Artin-Rees lemma. Krull's intersection theorem. Zariski rings

Let G a be set, which is at the same time an abelian group (written additively) and a topological space, not necessarily Hausdorff. We suppose the algebraic and topological structures above are compatible in the following sense: the mappings $G \times G \to G$ defined by $(x, y) \mapsto x + y$ and $G \to G$ defined by $x \mapsto -x$ are continuous (equivalently: the mapping $G \times G \to G$ given by $(x, y) \mapsto x - y$ is continuous). Then we say that G is a *topological group* (with respect to the two structures above).

Let G be a topological group and a an element of G. Then the translation T_a defined by $T_a(x) = x + a$ is continuous and the translation T_{-a} is its inverse mapping; so T_a is a homeomorphism of G into G and there is a bijective mapping of the set of all neighborhoods of 0 into the set of all neighborhoods of a $(U \mapsto U + a)$. Thus the topology of G is uniquely determined by the neighborhoods of 0 in G.

We may observe that every open subgroup H of G is closed; in fact the complementary set of H in G is $\bigcup_{x \notin H} x + H$ and every $x + H$ is open.

All properties of the following lemma are either trivial or follow easily by the continuity of the group operation.

Lemma 1.1. *Let H be the intersection of all neighborhoods of* 0 *in a topological group G. Then*

(i) *H is the closure* $\overline{\{0\}}$ of $\{0\}$.
(ii) *H is a subgroup of G.*
(iii) *G/H is Hausdorff.*
(iv) *G is Hausdorff* $\Leftrightarrow H=\{0\}$.

If A is a ring and a topological group with respect to addition, we say that A is a *topological ring* if multiplication (i.e. the mapping $A\times A\to A$ defined by $(x,y)\to(xy)$) is continuous.

If $\mathfrak{m}$ is an ideal of A, we may consider in A the topology defined by taking the set of all powers $\mathfrak{m}^n$ $(n\geqslant 0)$ as a fundamental system of neighborhoods of 0; it easy to check that with this topology A is a topological ring. We call this topology the $\mathfrak{m}$-*adic topology*, or simply the $\mathfrak{m}$-*topology*, and we say that A is an $\mathfrak{m}$-*adic ring*, or $\mathfrak{m}$-*ring*. Sometimes we write $(A,\mathfrak{m})$ instead of A, if we want to emphasize that we are considering the ring A provided with the $\mathfrak{m}$-topology. Notice that an $\mathfrak{m}$-ring A may be an $\mathfrak{m}'$-ring with $\mathfrak{m}\neq\mathfrak{m}'$. For instance, if A is a noetherian ring, it is easy to check that by taking $\mathfrak{m}'=\sqrt{\mathfrak{m}}$ (=radical of the ideal $\mathfrak{m}$), we have such a situation.

Likewise for an A-module E, where A is a topological ring and E a topological group: we say that E is a *topological module* if the mapping $A\times E\to E$ defined by $(a,x)\mapsto ax$ $(a\in A, x\in E)$ is continuous. The $\mathfrak{m}$-topology on E is defined by taking the $\mathfrak{m}$-topology on A and the topology given by all the submodules $\mathfrak{m}^n E$ $(n\geqslant 0)$ of E; E is really a topological module with respect to this topology.

If A is a topological ring such that a fundamental system of neighborhoods of 0 is given by ideals, we say that the topology on A is a *linear topology*. Therefore the $\mathfrak{m}$-topology is an example of linear topology.

Let E be an A-module. A chain $E=E_0\supset E_1\supset\cdots\supset E_n\supset\cdots$, where the E_n are submodules of E is called a *filtration* of E. A filtration is associated in a natural way with a topology in E: namely the one we get by taking $\{E_n\}$ as a fundamental system of neighborhoods of 0.

Let (E_n) be a filtration of E and $\mathfrak{m}$ an ideal of A such that the following condition is verified:

$$\mathfrak{m}E_n \subset E_{n+1} \quad \text{for all } n \geqslant 0, \text{ or equivalently:}$$
$$\mathfrak{m}^i E_n \subset E_{n+i} \quad \text{for all } n, i \geqslant 0;$$

then we call (E_n) an $\mathfrak{m}$-*filtration*. Thus the $\mathfrak{m}$-topology of E is finer than (but not necessarily equal to) the topology defined by any $\mathfrak{m}$-filtration of E.

We say that the $\mathfrak{m}$-filtration (E_n) is *stable* if there exists $n_0 \geqslant 0$ such that the following equivalent conditions are verified:

$$E_{n+1} = \mathfrak{m}E_n \qquad \text{for } n > n_0,$$
$$E_n = \mathfrak{m}^{n-n_0}E_{n_0} \quad \text{for } n > n_0,$$
$$E_{n+q} = \mathfrak{m}^q E_n \qquad \text{for } n > n_0,\ q \geqslant 0.$$

The filtration $(\mathfrak{m}^n E)$, which defines the $\mathfrak{m}$-adic topology in E, is obviously a stable $\mathfrak{m}$-filtration; conversely our next lemma shows that any stable $\mathfrak{m}$-filtration of M defines the $\mathfrak{m}$-topology.

Lemma 1.2. *If* $(E_n), (E'_n)$ *are stable* $\mathfrak{m}$*-filtrations of* E, *then they have bounded differences, that is, there exists* n_0 *such that* $E_{n+n_0} \subset E'_n$ *and* $E'_{n+n_0} \subset E_n$ *for all* $n \geqslant 0$. *Hence all stable* $\mathfrak{m}$*-filtrations determine the same topology on* M, *namely the* $\mathfrak{m}$*-topology* (for the easy proof of the lemma, see [3], lemma 10.6 or [6], p. 64, prop. 4).

Let A be a ring, and $\mathfrak{m}$ an ideal of A. Then we can form the graded ring $A' = \bigoplus_{n \geqslant 0} \mathfrak{m}^n$ (with multiplication defined as for polynomials) which is isomorphic to the graded subring $\sum_{n \geqslant 0} \mathfrak{m}^n X^n$ of the polynomial ring $A[X]$, where X is an indeterminate; A' is an A-algebra generated by $\mathfrak{m}X$, and so A' is noetherian whenever A is. Similarly if E is an A-module and (E_n) is an $\mathfrak{m}$-filtration of E, then $S' = \bigoplus_{n \geqslant 0} E_n$ is a graded A'-module. With this notation we have the following result, whose proof is not trivial at all (see, e.g., [6], p. 60, Th. 1).

Theorem 1.3. *Let* A *be a ring,* $\mathfrak{m}$ *an ideal of* A, (E_n) *an* $\mathfrak{m}$*-filtration of* E *such that* E_n *is a finitely generated submodule of* E.

Then the following conditions are equivalent:

a) *The filtration* (E_n) *is* $\mathfrak{m}$*-stable;*

b) E' *is a finitely generated* A'*-module.*

By Theorem 1.3 we get the following result, which is a very useful tool in our theory.

Theorem 1.4. (Artin-Rees lemma). *Let* A *be a noetherian ring,* $\mathfrak{m}$ *an ideal of* A, E *a finitely generated* A*-module,* (E_n) *a stable* $\mathfrak{m}$*-filtration of* E, F *a submodule of* E. *Then*

a) $(E_n \cap F)$ *is a stable* $\mathfrak{m}$*-filtration of* F.

b) *There exists an integer* n_0 *such that*

$$(\mathfrak{m}^n E) \cap F = \mathfrak{m}^{n-n_0}((\mathfrak{m}^{n_0} E) \cap F))$$

for all $n \geqslant n_0$.

c) *The filtrations* $(\mathfrak{m}^n F)$ and $((\mathfrak{m}^n E) \cap F)$ *have bounded differences; in particular the* $\mathfrak{m}$*-topology of* F *coincides with the topology induced on* F *by the* $\mathfrak{m}$*-topology of* E.

Proof. Since A in noetherian, F and each $F \cap E_n$ are finitely generated A-modules. Moreover $(E_n \cap F)$ is an $\mathfrak{m}$-filtration, since $\mathfrak{m}(E_n \cap F) \subset (\mathfrak{m} E_n) \cap (\mathfrak{m} F) \subset E_{n+1} \cap \mathfrak{m} F$, and defines a graded submodule $F' = \bigoplus_{n \geqslant 0} E_n \cap F$ of $E' = \bigoplus E_n$. Now E' is a finitely generated A'-module (Th. 1.3), and since A' is noetherian, F' is a finitely generated A'-module. Thus a) follows by Theorem 1.3. Moreover, if we apply a) to the filtration $(\mathfrak{m}^n E)$ of E, we get b), while c) follows by b) and Lemma 1.2. □

Remark 1. Statement b) is what is usually known as the Artin-Rees Lemma, while c) is a topological but weaker version of that result.

Remark 2. Theorem 1.4. may be false if A is not noetherian. There exist in fact a ring A with a unique maximal ideal $\mathfrak{m}$, and an element $a \in A$ such that: 1. A is $\mathfrak{m}$-Hausdorff; 2. if $E = aA$, b) is false ([6], p. 116, ex. 1).

In [2] it is shown that there also exist a ring B and an element $a \in B$ such that, if $A = B[[X]]$, $\mathfrak{m} = (X)$ and $E = (X - a)A$, the subset $\mathfrak{m} E$ is not open with respect to the induced topology, and thus c) is false in this case ([2], § 2). Notice that A is $\mathfrak{m}$-complete (see § 2), and that a can be chosen such that E is free.

Now we recall some standard definitions: a ring A is *local* (resp. *semilocal*) if it has only one maximal ideal (resp. a finite number of maximal ideals). The intersection of all maximal ideals of a ring A is the *(Jacobson) radical* of A and is written rad A.

We have $x \in \operatorname{rad} A$ if and only if every element of the set $1 - xA$ is a unit.

Theorem 1.5. (Krull's intersection theorem). *Let A be a noetherian ring, $\mathfrak{m}$ an ideal of A, E a finitely generated A-module; let $F = \bigcap \mathfrak{m}^n E$. Then $x \in F$ if and only if there exists $m \in \mathfrak{m}$ such that $(1-m)x = 0$.*

In particular we have $F = 0$, if at least one of the following hypotheses is verified:

a) $\mathfrak{m} \subset \operatorname{rad} A$.

b) *A is a local ring.*

c) *A is an integral domain.*

Proof. If $x = mx$, with $m \in \mathfrak{m}$, we have $x = \mathfrak{m}^n x$ for all $n \geqslant 0$, so $x \in F$. Reciprocally, if $x \in F$, Ax is contained in all neighborhoods of 0 in E, so it has no other open set than itself for the induced topology, which (by theor. 1.4.) is the $\mathfrak{m}$-topology of Ax; $\mathfrak{m}x$ is open for this topology, so $\mathfrak{m}x = Ax$ and $mx = x$ for some $m \in \mathfrak{m}$. The second part of the statement is trivial. ▯

Remark. Theorem 1.6 is false if A is not noetherian. As a counterexample take the ring A of C^∞ real functions of one variable, and the ideal $\mathfrak{m}$ of A consisting of all functions of A which vanish at the origin (for details see, e.g., [6], p. 65, Remarque, or [3], p. 110, Remark 2).

It is not hard to prove the following proposition (see, e.g., [6], p. 66).

Proposition 1.6. *Let A be a noetherian ring, $\mathfrak{m}$ an ideal of A. The following are equivalent:*

a) $\mathfrak{m} \subset \operatorname{rad} A$;

b) *every finitely generated A-module is* $\mathfrak{m}$-Hausdorff;

c) *if E is any finitely generated A-module, every submodule of E is closed for the $\mathfrak{m}$-topology of E;*

d) *every maximal ideal of A is closed for the $\mathfrak{m}$-topology.*

Definition. A noetherian topological ring is said to be a *Zariski ring* if it is an $\mathfrak{m}$-adic ring for some $\mathfrak{m}$ satisfying the equivalent conditions of Proposition 1.6.

Examples. If A is a local noetherian ring and $\mathfrak{m}$ is its maximal ideal we have $\mathfrak{m} = \operatorname{rad} A$, so $(A, \mathfrak{a})$ is a Zariski ring for any proper ideal $\mathfrak{a}$ of A.

If A is noetherian and complete for the $\mathfrak{m}$-topology, then $(A, \mathfrak{m})$ is a Zariski ring (see 2.18).

§ 2

Completions of filtered groups, rings and modules. Applications to $\mathfrak{m}$-adic topologies

Let G be a filtered abelian group (i.e. a filtered $\mathbb{Z}$-module, see § 1), and let (G_n) be its filtration. We can define a mapping

$$v: G \to \mathbb{N} \cup \{\infty\}$$

in the following way: $v(x) = \sup\{n \in \mathbb{N} \mid x \in G_n\}$. It is clear that $v(x) = \infty$ if and only if $x \in \bigcap G_n$, i.e. $x \in \{\overline{0}\}$ (lemma 1.1). The mapping allows us to define a *pseudometric* in G: let

$$d: G \times G \to G$$

be the mapping defined by $d(x, y) = e^{-v(x-y)}$ (we agree that $e^{-\infty} = 0$). Then it is easy to see that

$$d(x, y) \leqslant \sup\{d(x, z), d(y, z)\}$$

and that d defines in G the topology induced by the filtration (G_n).

Observe that d is a metric if and only if $\bigcap G_n = \{0\}$, i.e. if and only if G is Hausdorff (lemma 1.1). It is also clear that d induces a metric on $G/\bigcap G_n$.

Thus if G is a filtered group we can define limits and Cauchy sequences. We say that a sequence (x_n) of elements of G is *convergent* if there is an $x \in G$ such that $\lim d(x, x_n) = 0$; we say also that x is a limit for (x_n) and write: $x = \lim x_n$. A convergent sequence is always a *Cauchy sequence*, that is $\lim d(x_n, x_m) = 0$ as $m, n \to \infty$. The converse is not always true.

Definition 2.1. A filtered group G is *complete* if and only if the following equivalent conditions are satisfied:

(i) G is Hausdorff and any Cauchy sequence of elements of G is convergent;

(ii) Every Cauchy sequence of elements of G has a unique limit (in G).

Now we want to associate with any filtered group a complete one in a natural way. We begin with some preliminaries.

Definition 2.2. A continuous homomorphism $f: G \to G'$ of topological groups is said to be *strict* if the quotient topology of $f(G)$ coincides with the topology induced by G'.

Note that if G' is any non discrete topological group, and G is the same group but with the discrete topology, the identity map $G \to G'$ is continuous but not strict.

Our next proposition gives two equivalent conditions we shall use to define a completion.

Proposition 2.3. *Let G be a filtered group, (G_n) its filtration, G' a complete filtered group and $f: G \to G'$ a continuous homomorphism. Then the following conditions are equivalent:*

(i) $\operatorname{Ker} f = \bigcap G_n$, *$f$ is strict and $f(G)$ is dense in G';*

(ii) *for any complete filtered group H and any continuous homomorphism $g: G \to H$ there is a unique continuous homomorphism $g': G' \to H$ such that $g' \circ f = g$.*

Proof. We show first that (i)⇒(ii). Since H is Hausdorff, $\operatorname{Ker} g$ is closed and hence contains the closure of $0 \in G$, which coincides with $\bigcap G_n$ (Lemma 1.1). Thus g factors uniquely through $f(G)$, and since f is strict, we may suppose $G = f(G)$. In order to get g' we extend g in the following way. Let $x \in G'$ and let $x = \lim x_n$, where (x_n) is a (Cauchy) sequence of elements of G. Then $(g(x_n))$ is a Cauchy sequence of elements of H, which has a unique limit y, since H is complete. It is easy to show that y depends only on x (and not on the sequence (x_n)), and that the mapping $g': x \to y$ is a group homomorphism. Moreover g' is continuous and unique by well known topological facts.

We want now to prove that (ii)⇒(i). We show first that if (ii) holds, then $\operatorname{Ker} f = \bigcap G_n$. Since f is continuous and G' is Hausdorff, we have $\operatorname{Ker} f \supset \bigcap G_n$. Conversely, let $x \in \operatorname{Ker} f$ and let n be an integer. Since the quotient group G/G_n is discrete, and hence complete, the canonical map $\pi_n: G \to G/G_n$ factors uniquely through a continuous homomorphism $\pi'_n: G' \to G/G_n$. Therefore we have

$$\pi_n(x) = \pi'_n(f(x)) = \pi'(0) = 0$$

whence $x \in G_n$. Thus $\operatorname{Ker} f \subset \bigcap G_n$.

We prove now that f is strict, i.e. that the topology induced by G' on $f(G)$ coincides with the topology associated with the filtration $(f(G_n))$. With the above notations we have $f(G_n) = f(G) \cap \operatorname{Ker} \pi'_n$, where $\operatorname{Ker} \pi'_n$ is open in G' since G/G_n is discrete. Thus the topology induced by G' on $f(G)$ is finer than the quotient topology; and since f is continuous, it is easy to see that the quotient topology is finer than the induced one: then these topologies must coincide.

To conclude the proof we have to show that $f(G)$ is dense in G', i.e. that the closure G'' of $f(G)$ in G' coincides with G'. Since G' is complete, so is G'', and hence there is a unique continuous homomorphism $g': G' \to G''$ such that $g' \circ f = g'$. It is easy to see that if $i: G'' \to G'$ is the inclusion, then $i \circ g'$ is the identity map of G'; thus i is surjective and the proof is complete. ▯

Definition 2.4. Let G be a filtered group. A *completion* of G is a pair (G', f), where G' is a complete filtered group and $f: G \to G'$ is a continuous homomorphism, satisfying the equivalent conditions of Proposition 2.3. We say also that G' is a completion of G, if the map f is understood.

Before we show that a completion exists, we give some easy consequences of the above results.

Corollary 2.5. *Let (G', f) be a completion of the filtered group G. Then*

(i) *f is injective if and only if G is Hausdorff.*

(ii) *f is bijective if and only if G is complete.*

Proof. It follows easily by Proposition 2.3 and Lemma 1.1. ▯

Corollary 2.6. *Let $u: G_1 \to G_2$ be a continuous homomorphism of filtered groups, and let (G'_i, f_i) be a completion of G_i $(i=1,2)$. Then there is a unique continuous homomorphism $u': G'_1 \to G'_2$ such that $u' \circ f_1 = f_2 \circ u$. Moreover if u is the identity so is u', and if $v: G_2 \to G_3$ is a second continuous homomorphism of filtered groups one has: $(v \circ u)' = v' \circ u'$.*

Proof. To show the first assertion apply condition (ii) of Proposition 2.3 with $G = G_1$, $H = G'_2$ and $g = f_2 \circ u$. The second is trivial, and to prove the third observe that if (G'_3, f'_3) is a completion of G_3, one has $(v \circ u)' \circ f_1 = f_3 \circ (v \circ u) = (v' \circ u') \circ f_1$ and the conclusion follows again by Proposition 2.3, ii). □

Our next corollary shows that *completion is unique (up to isomorphisms).*

Corollary 2.7. *Let G be a filtered group, and (G', f') (G'', f'') two completions of G. Then there is a unique isomorphism $u: G' \to G''$, which is also a homeomorphism, such that $u \circ f = f''$.*

Proof. It follows easily by Corollary 2.6. □

Now we want to show that any filtered group has a completion. Here we give the classical construction, based upon Cauchy sequences. Let $\mathcal{G}$ be the set of all Cauchy sequences of elements of G, and let $\mathcal{N}$ be the set of the zero-sequences, i.e. of sequences having 0 as a limit. Clearly $\mathcal{G}$ is a group, and $\mathcal{N}$ is a subgroup of $\mathcal{G}$. The quotient group $\hat{G} = \mathcal{G}/\mathcal{N}$ can be given a metric by setting

$$d(\overline{(x_n)}, \overline{(y_n)}) = \inf d(x_n, y_n)$$

where the upper bar denotes reduction modulo $\mathcal{N}$. It is not difficult to see that such a d is a well defined metric on $\hat{G}$, and that $\hat{G}$ is a topological group with respect to d.

Let now $f: G \to \hat{G}$ be the canonical map, which sends each $x \in G$ to the class of the constant sequence $(x, x, \ldots)$. We claim that $(\hat{G}, f)$ is a completion of G. More precisely we have:

Proposition 2.8. *Let G be a filtered group, (G_n) its filtration and $f: G \to \hat{G}$ the canonical map. Then*

(i) *$\hat{G}_n$ can be canonically identified with a subgroup of $\hat{G}$* $(n = 0, 1, \ldots)$.

(ii) *The topology of G is induced by the filtration* $(\hat{G}_n)$.

(iii) *The filtered group $\hat{G}$, together with the map f, is a completion of G.*

Proof. With the above notations we have $\hat{G}_n = \mathcal{G}_n/\mathcal{N}_n$ for any n. It is easy to see that $\mathcal{N}_n = \mathcal{N} \cap \mathcal{G}_n$, whence $\hat{G}_n \cong (\mathcal{G}_n + \mathcal{N})/\mathcal{N}$. This proves (i). By the definition of the metric in $\hat{G}$ we have: $\alpha \in \hat{G}_n$ if and only if $d(\alpha, 0) \leqslant e^{-n}$; this shows that $(\hat{G}_n)$ is a fundamental system of neighbourhoods of 0, and (ii) follows.

To prove (iii) we use condition (i) of proposition 2.3. First of all we note that, as a metric space, $\hat{G}$ is just the completion of the metric space $G/\bigcap G_n$, and then $\hat{G}$ is complete. By (ii) it follows easily that f is strict. Moreover $f(G)$ is just the image of $G/\bigcap G_n$ in $\hat{G}$, and this shows that $\operatorname{Ker} f=\bigcap G_n$, and that $f(G)$ is dense in $\hat{G}$. Thus $(\hat{G},f)$ is a completion of G by Proposition 2.3, (i). □

Now we want to show that the completion of a filtered group can be obtained by means of *inverse limits*. We recall that a sequence (A_n, f_n) of abelian groups A_n and group homomorphisms $f_n: A_{n+1}\to A_n$ is called an *inverse system* and has an *inverse limit* $\varprojlim A_n = B$ defined by

$$B=\left\{(a_n)\in\prod A_n \,\middle|\, f_n(a_{n+1})=a_n \text{ for any } n\right\}.$$

This is clearly a subgroup of $\prod A_n$, and for any n there is a canonical homomorphism $g_n: B\to A_n$ such that $g_{n+1}\circ f_n = g_n$ for any n.

If G is a filtered group and (G_n) is its filtration, we have the inverse systems $(G/G_n, f_n)$ and, for fixed i, $(G_n/G_{n+i}, f_{n+i})$, where f_n and f_{n+i} are the canonical mappings. Let $\tilde{G}=\varprojlim G/G_n$, $\tilde{G}_n=\varprojlim G_n/G_{n+i}$. It is easy to see that $(\tilde{G}_n)$ is a filtration of $\tilde{G}$. Note finally that there is an obvious homomorphism $g: G\to\tilde{G}$ defined by $f(x)=(x_0, x_1, \ldots)$ where x_i is the class of $x\in G$ modulo G_i.

Proposition 2.9. *The filtered group $\tilde{G}$, together with the homomorphism $g: G\to\tilde{G}$, is a completion of G.*

Proof. It is sufficient to show that there is a canonical isomorphism $u:\hat{G}\to\tilde{G}$ such that

(i) $\qquad u\circ f = g,$

(ii) $\qquad u(\hat{G}_n)=\tilde{G}_n \quad \text{for any } n=0,1,\ldots.$

Let then (x_n) be a Cauchy sequence of elements of G. Then for any r there is an n such that $x_n\equiv x_{n+1}\equiv x_{n+2}\equiv\cdots$ modulo G_r. Denote by ξ_r the class of those x_{n+i}'s. We get a sequence $\xi=(\xi_0,\xi_1,\ldots)$ which is clearly an element of $\tilde{G}$. This gives a group homomorphism $\mathscr{G}\to\tilde{G}$, whose kernel is clearly $\mathscr{N}$. Thus we get a canonical injective homomorphism $u:\hat{G}\to\tilde{G}$. Moreover if $\xi=(\xi_n)\in\tilde{G}$ and $x_n\in G$ is a lifting of $\xi_n\in G/G_n$, it is easy to see that (x_n) is a Cauchy sequence of elements of G, and that ξ is the image under u, of the class of (x_n). Thus u is surjective, and (i) is proved. A similar argument shows (ii), and the proof is complete. □

Proposition 2.9 is useful in the proof of the following proposition (see e.g. [3], cor. 10.3).

Proposition 2.10. *Let* $0 \to G' \to G \to G'' \to 0$ *be an exact sequence of filtered groups and strict homomorphisms. Then the corresponding sequence* $0 \to \hat{G}' \to \hat{G} \to \hat{G}'' \to 0$ (cor. 2.6) *is exact.*

Corollary 2.11. *Let G be a filtered group and (G_n) its filtration. Then the canonical mapping $f: G \to \hat{G}$ induces isomorphisms:*

(i) $\quad f_n: G/G_n \to \hat{G}/\hat{G}_n \qquad (n = 0, 1, \ldots)$,

(ii) $\quad f_{n,i}: G_n/G_{n+1} \to \hat{G}_n/\hat{G}_{n+i} \qquad (n, i = 0, 1, \ldots)$.

Proof. Since G/G_n is discrete and hence complete, conclusion (i) follows by Proposition 2.10 applied to the exact sequence $0 \to G_n \to G \to G/G_n \to 0$. The proof of (ii) is similar. □

The above results can be easily extended to filtered rings and modules: it is in fact a routine business to verify that ring and module operations can be naturally defined, and work as they are expected to do. Here we list some results.

Proposition 2.12. *Let A be a filtered ring, $(\mathfrak{a}_n)$ its filtration and $(\hat{A}, f)$ the completion of A as a filtered group. Then:*

a) *$\hat{A}$ can be given a unique topological ring structure, such that f is a continuous ring homomorphism.*

b) *The completion $\hat{\mathfrak{a}}_n$ of each $\mathfrak{a}_n$ can be embedded in $\hat{A}$ as an ideal, and the topology of $\hat{A}$ is induced by the filtration $(\hat{\mathfrak{a}}_n)$.*

Proposition 2.13. *Let A be a filtered ring, M a filtered A-module and (M_n) its filtration. Let $\hat{A}$ be the completion of A* (Prop. 2.12), *and $\hat{M}, \hat{M}_n$ the completions of M and M_n as filtered groups. Then, if M is a topological A-module* (§ 1) *we have:*

a) *$\hat{M}$ has a unique topological $\hat{A}$-module structure, such that the canonical homomorphism $M \to \hat{M}$ is A-linear.*

b) *Each $\hat{M}_n$ is canonically embedded in $\hat{M}$ as a sub-$\hat{A}$-module, and the topology of $\hat{M}$ is induced by the filtration $(\hat{M}_n)$.*

If A is a filtered ring and $(\mathfrak{a}_n)$ is its filtration, we can define the associated graded ring $Gr(A) = \oplus \mathfrak{a}_n/\mathfrak{a}_{n+1}$, with multiplication as for polynomials. Likewise if M is an A-module and (M_n) is a filtration of M such that $\mathfrak{a}_n M \subset M_n$ for any n, we have the graded $Gr(A)$-module $Gr(M) = \oplus M_n/M_{n+1}$. Then by Corollary 2.11 we get easily:

Corollary 2.14. *There exist a canonical isomorphism of graded rings $Gr(A) \cong Gr(\hat{A})$ and a canonical isomorphism of graded $Gr(A)$-modules $Gr(M) \cong Gr(\hat{M})$.*

If A is an $\mathfrak{m}$-adic ring (§ 1) the completion of A is called the $\mathfrak{m}$-completion of A, and is denoted by $(A, \mathfrak{m})\hat{}$ or simply by $\hat{A}$ if $\mathfrak{m}$ is understood. Likewise if M is an A-module we have the $\mathfrak{m}$-completion of M, denoted by $(M, \mathfrak{m})\hat{}$ or simply by $\hat{M}$.

Remark. It is not always true that the $\mathfrak{m}$-completion $\hat{A}$ of the ring A is an $\hat{\mathfrak{m}}$-adic ring. It might even happen that $\hat{A}$ is not $\hat{\mathfrak{m}}$-complete (see e.g. [6], p. 107, ex. 2).

We shall see, however, that if $\mathfrak{m}$ is finitely generated, then $\hat{\mathfrak{m}} = \mathfrak{m}\hat{A}$, and $\hat{A}$ has the $\hat{\mathfrak{m}}$-topology (see 4.3).

Clearly one can apply to $\mathfrak{m}$-completions all the results valid for filtered rings. In order to have a quick reference for the future, we summarize the main results we shall need later.

Proposition 2.15. *Let A be an $\mathfrak{m}$-adic ring and M an A-module. Then:*

a) *The $\mathfrak{m}$-completions $\hat{A}$ and $\hat{M}$ exist, and $\hat{M}$ is a topological $\hat{A}$-module.*

b) $\hat{A} \cong \varprojlim A/\mathfrak{m}^n$ and $\hat{M} \cong \varprojlim M/\mathfrak{m}^n M$.

c) *The graded rings $Gr(A)$ and $Gr(\hat{A})$ are isomorphic (in particular $A/\mathfrak{m} \cong \hat{A}/\hat{\mathfrak{m}}$).*

d) *The graded modules $Gr(M)$ and $Gr(\hat{M})$ are isomorphic (in particular the $(A/\mathfrak{m})$-modules $M/\mathfrak{m}M$ and $\hat{M}/(\mathfrak{m}M)\hat{}$ are isomorphic).*

Our next propositions describe the "functorial" properties of $\mathfrak{m}$-completions.

Proposition 2.16. *Let A be an $\mathfrak{m}$-adic ring, B an $\mathfrak{n}$-adic ring and $f: A \to B$ a homomorphism. Suppose there is an $n > 0$ such* *that* $f^{-1}(\mathfrak{n}) \supset \mathfrak{m}^n$. *Then*

a) f *is continuous.*

b) f *induces a unique continuous homomorphism* $\hat{f}: \hat{A} \to \hat{B}$.

Proof. a) follows by an easy verification, and b) is a consequence of a) and of Corollary 2.6. □

Proposition 2.17. *Let* A *be an* $\mathfrak{m}$*-adic ring,* M, N *two* A*-modules and* $f: M \to N$ *a homomorphism. Then* f *induces a unique continuous homomorphism* $\hat{f}: \hat{M} \to \hat{N}$ *of* $\hat{A}$ *modules. If moreover* f *is surjective, then* $\hat{f}$ *is surjective.*

Proof. Since f is clearly $\mathfrak{m}$-continuous $\hat{f}$ exists and is unique by Corollary 2.6.

Moreover it is easy to see that if f is surjective, it is also strict with respect to the $\mathfrak{m}$-topologies. Then $\hat{f}$ is surjective by Proposition 2.10. □

Now we want to study more closely the ideal $\hat{\mathfrak{m}}$.

Corollary 2.18. *Let* A *be an* $\mathfrak{m}$*-adic ring, and* $\hat{A}$ *its completion. Then:*

a) *For any* $x \in \hat{\mathfrak{m}}$, *one has:* $\lim x^n = 0$.

b) $\hat{\mathfrak{m}} \subset \operatorname{rad} \hat{A}$.

Proof. By Proposition 1.15 d) one has $\mathfrak{m}/\mathfrak{m}^n = \hat{\mathfrak{m}}/(\mathfrak{m}^n)\hat{}$ for any n. Thus the class of x modulo $(\mathfrak{m}^n)\hat{}$ is nilpotent. This shows a). It follow that the series $1 + x + x^2 + \cdots$ converges in $\hat{A}$, so $1 - x$ is invertible, and this proves b). □

By Corollary 2.18 and the isomorphism $A/\mathfrak{m} = \hat{A}/\hat{\mathfrak{m}}$ we get

Corollary 2.19. *Let* A *be an* $\mathfrak{m}$*-adic ring,* $\hat{A}$ *its completion and* $f: A \to \hat{A}$ *the canonical homomorphism. Then the mapping* $\mathfrak{m} \mapsto f^{-1}(\mathfrak{m})$ *is a bijection of the set of the maximal ideals of* $\hat{A}$ *onto the set of those maximal ideals of* A *which contain* $\mathfrak{m}$, *the inverse map being* $\mathfrak{n} \mapsto \hat{\mathfrak{n}}$.

Corollary 2.20. *Let* A *be a local ring and* $\mathfrak{m}$ *a maximal ideal of* A. *Then the* $\mathfrak{m}$*-completion of* A *is a local ring with maximal ideal* $\hat{\mathfrak{m}}$.

If A is a local ring, we shall often consider in A the $\mathfrak{m}$-topology, where $\mathfrak{m}$ in the maximal ideal of A.

In this case $\mathfrak{m}$ is often understood, and we shall say, for example, *completion* instead of $\mathfrak{m}$-completion, *complete* instead of $\mathfrak{m}$-complete, etc.

Note, however, that with this terminology, the completion of a local ring (which is a local ring by Corollary 2.20), is not necessarily complete, since we have seen that the topology of an $\mathfrak{m}$-completion is not necessarily $\hat{\mathfrak{m}}$-complete.

We recall that if A and B are local rings with maximal ideals $\mathfrak{m}$ and $\mathfrak{n}$ respectively, a homomorphism $f: A \to B$ is said to be *local* if $f^{-1}(\mathfrak{n}) = \mathfrak{m}$.

By Corollary 2.20 and Proposition 2.16 we have:

Corollary 2.21. *Let $f: A \to B$ be a local homomorphism of local rings. Then f induces a unique local homomorphism $\hat{f}: \hat{A} \to \hat{B}$.*

We conclude this section with

Proposition 2.22. *Let A be a ring, $\mathfrak{m}$ an ideal of A, S the multiplicative set $1+\mathfrak{m}$, and $B = S^{-1}A$. Then $(B, \mathfrak{m}B)\hat{} = (A, \mathfrak{m})\hat{}$*

Proof. It is not difficult to see that $B/(\mathfrak{m}B)^n \cong A/\mathfrak{m}^n$ for any n. The conclusion follows then by Proposition 2.15 b). ◻

Remark 1. If A is a noetherian ring, the ring B of Proposition 2.22 is a Zariski ring with respect to the $\mathfrak{m}B$-topology (cfr. prop. 1.6), and it is often called the "zariskification" of A with respect to $\mathfrak{m}$. Thus a completion of a noetherian ring can be regarded as the completion of its zariskification. This may have several advantages, as we will see later.

Remark 2. Our treatment of completions has been limited to filtered groups and rings, a more general treatment being beyond our purpose. For further reading on this subject we refer to [9], [10] and [6].

§ 3

Rings of formal and restricted power series. Preparation theorems. Hensel lemma

Let $A[[X_1, \ldots, X_n]]$ be the ring of formal power series in n indeterminates over the ring A; the elements of $A[[X_1, \ldots, X_n]]$ are the series $f = f_0 + f_1 + \cdots + f_n + \cdots$ where each f_i is a homogeneous polynomial of degree i in $X_1, \ldots, X_n$. The ring $A[[X_1, \ldots, X_n]]$ is complete for the $(X_1, \ldots, X_n)$-topology. In fact, if (g_n) $(g_n \in A[X_1, \ldots, X_n])$ is a Cauchy sequence, the series $g_0 + (g_1 - g_0) + \cdots + (g_{n+1} - g_n) + \cdots \in A[[X_1, \ldots, X_n]]$ is the limit of (g_n); furthermore $A[[X_1, \ldots, X_n]]$ is Hausdorff, as $\bigcap_m (X_1, \ldots, X_n)^m = 0$. The $(X_1, \ldots, X_n)$-adic topology of $A[[X_1, \ldots, X_n]]$ induces the $(X_1, \ldots, X_n)$-topology on the polynomial ring $A[X_1, \ldots, X_n]$ which is a subring of $A[[X_1, \ldots, X_n]]$; every series may be approximated by polynomials, so the ring $A[X_1, \ldots, X_n]$ is dense in $A[[X_1, \ldots, X_n]]$. Thus we have the following proposition.

Proposition 3.1. *The ring* $A[[X_1, \ldots, X_n]]$ *is the completion of the ring* $A[X_1, \ldots, X_n]$ *for the* $(X_1, \ldots, X_n)$*-topology.*

Our next theorem corresponds to the Hilbert basis theorem for the case of formal power series.

Theorem 3.2. *If* A *is noetherian, then* $A[[X_1, \ldots, X_n]]$ *is also noetherian.*

We may give direct proofs of Theorem 3.2 (see, e.g. [41], p. 138, Th. 4 or [29], 15.3, p. 50) or we may deduce it as a particular case of a general property about completions of noetherian rings (see § 5).

The ring of formal power series is very important in studying completions, because of this fact: if $\mathfrak{m} = (a_1, \ldots, a_n)$ is a finitely generated ideal of the ring A and $\hat{A}$ is the $\mathfrak{m}$-completion of A, there exists a surjective homomorphism $\varphi\colon A[[X_1,\ldots,X_n]]\to\hat{A}$ such that $\varphi(X_i)=a_i$ $(1\leqslant i\leqslant n)$ and $\operatorname{Ker}\varphi$ is the $(X_1,\ldots,X_n)$-closure in $A[[X_1,\ldots,X_n]]$ of the ideal $(X_1-a_1,\ldots,X_n-a_n)$ (see, e.g. [29], 17.5. p. 50, or [2], n^0 1, or also [6], p. 114, ex. 26, where it is shown that the above result may be generalized to ideals with any number of generators). If A is noetherian, $A[[X_1,\ldots,X_n]]$ is clearly a Zariski ring for the $(X_1,\ldots,X_n)$-topology, as a consequence of 3.2 and 1.6, so every ideal is closed in $A[[X_1,\ldots,X_n]]$. Therefore we have the following result.

Proposition 3.3. *If A is a noetherian ring, and $\mathfrak{m}=(a_1,\ldots,a_n)$ is a finitely generated ideal of A such that A is Hausdorff for the $\mathfrak{m}$-topology, then the $\mathfrak{m}$-completion of A is isomorphic to the ring $A[[X_1,\ldots,X_n]]/(X_1-a_1,\ldots,X_n-a_n)$.*

The above result is not true in general for $\mathfrak{m}$-completions of finitely generated ideals in a non-noetherian ring A; in [2], n^{0}2 we may find a counterexample and also some sufficient conditions for the isomorphism above, which are very simple in the case of a principal ideal as is shown by the next proposition.

Proposition 3.4. *If (a) is a principal ideal of the ring A, and the sequence*

$$\operatorname{Ann}(a)\subset\operatorname{Ann}(a^2)\subset\cdots\subset\operatorname{Ann}(a^n)\subset\cdots$$

is stationary, the (a)-completion of A is isomorphic to the ring $A[[X]]/(X-a)$.

Now we consider one of the several important properties related to complete local rings. In fact, for formal power series over complete local rings there exists a "preparation theorem": in order to present it, some definitions are useful.

Let A be a local ring, $\mathfrak{m}$ its maximal ideal. We say that the series $f=\sum_0^\infty a_i X^i\in A[[X]]$ is *regular* if some coefficient a_i is a unit; if f is regular and s is the smallest integer such that a_i is a unit, we say that f has *order* s. If $p=a_0+a_1X+\cdots+a_{t-1}X^{t-1}+X^t$ is a monic polynomial of degree t and all coefficients $a_0,a_1,\ldots,a_{t-1}$ are in $\mathfrak{m}$, we say that p is a *distinguished polynomial* of degree t.

Theorem 3.5. (Preparation theorem). *Let A be a complete local ring, $\mathfrak{m}$ its maximal ideal, f a regular series of $A[[X]]$ of order s. Then we have:*

a) *if g is any series of $A[[X]]$, there exist a series $h \in A[[X]]$ and a polynomial $q \in A[X]$ of degree s, such that $g = fh + r$; furthermore q and r are uniquely determined;*

b) *there exist a unit $u \in A[[X]]$ and a distinguished polynomial $p \in A[X]$ of degree s such that $f = up$; u and p are uniquely determined.*

The proof of the above theorem may be found in [8], § 3, n° 9. □

Remark. If k is a field and $A = k[[X_1, \ldots, X_{n-1}]]$, Theorem 3.5. gives the classical Weierstrass preparation theorem for formal power series. There exists also a Weierstrass preparation theorem for convergent power series. Namely, if $k = \mathbb{R}$ or $k = \mathbb{C}$ (or, more generally a complete valued field), we may in 3.5 substitute for A the ring $k\{X_1, \ldots, X_{n-1}\}$ of convergent power series (in a neighborhood of the origin) and $A\{X\}$ with $k\{X_1, \ldots, X_n\}$; the proof in this case is more complicated, as we have in addition to verify that some formal series are convergent (see, e.g. [41], p. 142, or [29], 45.3).

We now consider in a ring A a linear topology τ (see § 1), which is not necessarily an $\mathfrak{m}$-topology; we assume for simplicity that there exists a countable fundamental system of neighborhoods given by the ideals $\mathfrak{m}_1, \mathfrak{m}_2, \ldots, \mathfrak{m}_n, \ldots$ so that the topology is given by the filtration

$$\mathfrak{m}_1, \mathfrak{m}_1 \cap \mathfrak{m}_2, \mathfrak{m}_1 \cap \mathfrak{m}_2 \cap \mathfrak{m}_3, \ldots$$

We say that a formal power series $f = \sum_0^\infty a_n X^n$ of $A[[X]]$ is restricted (with respect to τ) if $\lim_{n \to \infty} a_n = 0$. The definition may be generalised to the case of a finite number of indeterminates: a series $f = \sum a_{i_1 \ldots i_n} X_1^{i_1} \ldots X_n^{i_n}$ is called *restricted* if $\lim a_{i_1 \ldots i_n} = 0$ for $\min(i_1, \ldots, i_n) \to \infty$. It is easy to see that restricted power series are a subring of $A[[X_1, \ldots, X_n]]$ (see [6], § 4, n° 2, or [20], Ch. 0, 7.5). We denote this ring $A\{X_1, \ldots, X_n\}$ if there is no confusion about the linear topology we are considering.

We denote by M_i $(i \in \mathbb{N})$ the ideal of restricted series of $A\{X_1, \ldots, X_n\}$ whose coefficients are in $\mathfrak{m}_i$. Let σ be the linear topology of $A\{X_1, \dots, X_n\}$ given by the ideals $M_i(X_1, \dots, X_n)^m$ $(i, m \in \mathbb{N})$ and σ' be the topology induced by σ on the polynomial ring $A[X_1, \dots, X_n]$. Then it is easy to prove the following proposition (see [30], prop. 1).

Proposition 3.6. *The ring $A\{X_1, \dots, X_n\}$ is (Hausdorff and) complete with respect to the σ-topology; namely, it is the completion of the ring $A\{X_1, \dots, X_n\}$ provided with the σ'topology.*

In particular, if σ is an $\mathfrak{m}$-topology, then $A\{X_1, \dots, X_n\}$ is the completion of $A\{X_1, \dots, X_n\}$ with respect to the $\mathfrak{m}(X_1, \dots, X_n)$-topology.

We say that an element a of a topological ring A is *topologically nilpotent* if $\lim_{n \to \infty} a^n = 0$.

Theorem 3.7. (Hensel). *Let A be a complete Hausdorff ring with respect to a linear topology, $\mathfrak{m}$ a closed ideal whose elements are topologically nilpotent, $B = A/\mathfrak{m}$ the topological quotient ring, $\varphi: A\{X\} \to B\{X\}$ the canonical mapping. Let $f \in A\{X\}$, $\bar{p} \in B[X]$, $\bar{q} \in B\{X\}$ be given such that: $\varphi(f) = \bar{p}\bar{q}$, $\bar{p}$ is monic and $\bar{p}, \bar{q}$ are coprime in $B[X]$. Then there exist $p \in A[X]$, $q \in A\{X\}$ such that: $f = pq$, p is monic, $\varphi(p) = \bar{p}, \varphi(q) = \bar{q}$.*

Moreover p and q are uniquely determined and coprime in $A\{X\}$; if in addition, f is a polynomial, so is q.

The proof of Theorem 3.6. is given in [6], p. 84. □

In the classical "Hensel lemma" A is the ring $\mathbb{Z}_p$ of p-adic integers (that is the completion of the local ring $\mathbb{Z}_{(p)}$ for the (p)-topology), the topology in $\mathbb{Z}_p$ is the (p)-topology and $\mathfrak{m} = (p)$; furthermore in the classical statement there are no restricted series, but only polynomials. If A is a complete local ring, $\mathfrak{m}$ its maximal ideal and A is provided with the $\mathfrak{m}$-topology, the situation is essentially the same as in the case of p-adic integers. The generalisation stated in Theorem 3.6 is useful for some applications, like the following "preparation theorem" for restricted power series; the theorem is given directly for n indeterminates.

Theorem 3.8. *Let A, $\mathfrak{m}$, B be as in the hypothesis of Theorem 3.7. Let $\varphi: A\{X_1, \dots, X_n\} \to (A/\mathfrak{m})\{X_1, \dots, X_n\}$ be the canonical mapping and $f \in A\{X_1, \dots, X_n\}$ a series such that $\varphi(f)$ is a monic* *polynomial of degree s of the ring* $A\{X_1,\ldots,X_{n-1}\}[X_n]$. *Then we have:*

a) *every series g of* $A\{X_1,\ldots,X_n\}$ *may be written in one and only one way in the form* $g=fq+r$, *where* $q\in A\{X_1,\ldots,X_n\}$ *and r is a polynomial of* $A\{X_1,\ldots,X_{n-1}\}[X_n]$ *of degree* $<s$.

b) *g is associated in* $A\{X_1,\ldots,X_n\}$ *with one and only one polynomial* $p\in A\{X_1,\ldots,X_{n-1}\}[X_n]$ *monic in* X_n *and p has degree s* (see, for the proof, [30], Th. 10 and Cor. 1 to Th. 11). ▯

Remark. Let k be a complete valued non archimedean field, A the valuation ring of k, $\mathfrak{m}$ the maximal ideal of A; we consider on A the topology induced by the topology of k. Then we may apply Theorem 3.8 to get, in this way, a preparation theorem for the ring $k\{X_1,\ldots,X_n\}$ of convergent power series in the unit polydisc of k^n: see [31], Teor. 1. This result was also proved independently by Grauert and Remmert who developped with many others the theory of affinoid rings started by Tate in [40].

§ 4
Completions of finitely generated modules. Flatness and faithful flatness

In this paragraph we shall consider essentially completions of finitely generated modules, in particular finitely generated modules over a noetherian ring A.

Lemma 4.1. *Any* $\mathfrak{m}$*-completion commutes (up to isomorphisms) with finite direct sums. In particular we have* $\hat{A^n} \cong (\hat{A})^n$.

Proof. It follows immediately from Proposition 2.10. □

Theorem 4.2. *Let A be a ring, $\mathfrak{m}$ an ideal of A, M a finitely generated A-module, $\hat{A}$ and $\hat{M}$ the $\mathfrak{m}$-completions of A and M respectively. Then the canonical mapping $\varphi : \hat{A} \otimes_A M \to \hat{M}$ is surjective; equivalently: $\hat{M} = \hat{A} M$.*

Proof. Let $\beta : L \to M$ be a surjective homomorphism with L free and finitely generated. In the commutative diagram

$$\begin{array}{ccc} \hat{A} \otimes_A L & \xrightarrow{\alpha} & \hat{A} \otimes_A M \\ \downarrow{\scriptstyle \psi} & & \downarrow{\scriptstyle \varphi} \\ \hat{L} & \xrightarrow{\beta} & \hat{M} \end{array}$$

α is surjective (as the functor $M \mapsto \hat{A} \otimes_A M$ is right exact), ψ is an isomorphism by lemma 4.1 and β is surjective by Proposition 2.17. Then it follows that φ is surjective, that is to say $\hat{A} M = \operatorname{Im} \varphi = \hat{M}$. □

Corollary 4.3. *If $\mathfrak{m}$ is a finitely generated ideal of the ring A, then:*

(i) $\hat{\mathfrak{m}} = \hat{A} \mathfrak{m}$, *and more generally:*

(ii) $(\mathfrak{m}^n)\hat{} = (\hat{\mathfrak{m}})^n = \hat{A} \mathfrak{m}^n$ *for all* $n \geq 0$.

In particular the topology of $\hat{A}$ is the $\hat{\mathfrak{m}}$-topology and also the $\mathfrak{m}$-topology considering $\hat{A}$ as an A-module.

Proof. (i) follows from 4.2. Now apply (i) to $\mathfrak{m}^n$ and we deduce that

$$(\mathfrak{m}^n)\hat{} = \hat{A}\,\mathfrak{m}^n = (\hat{A}\,\mathfrak{m})^n = (\hat{\mathfrak{m}})^n .$$

Here we emphasize that, with the notation of theorem 4.2., $(\mathfrak{m}^n)\hat{}$ is the $\mathfrak{m}$-completion of $\mathfrak{m}^n$. Nevertheless the $\mathfrak{m}$-topology of $\mathfrak{m}^n$ is obviously induced by the $\mathfrak{m}$-topology of A, so the topology of $\hat{A}$ is given by the filtration $(\mathfrak{m}^n)\hat{}$ (by 2.8.). Then the last statement follows by (ii). □

Remark. Corollary 4.3. is also an immediate consequence of the following proposition which is not related to completions of finitely generated modules but to $\mathfrak{m}$-completions of any module when $\mathfrak{m}$ is finitely generated.

Proposition 4.4. *If $\mathfrak{m}$ is a finetely generated ideal of a ring A and $\hat{M}$ is the $\mathfrak{m}$-completion of a module M, then the closure of $\mathfrak{m}^n M$ in $\hat{M}$ is $\mathfrak{m}^n \hat{M}$ (for all $n \geqslant 0$).*

Proof. See the proof of Theorem 17.4 of [29]. □

In the following statements the additional hypothesis of rings being noetherian is fundamental.

Theorem 4.5. *Let $\mathfrak{m}$ be an ideal of a noetherian ring A. If*

$$0 \to M' \to M \to M'' \to 0$$

is an exact sequence of finitely generated A-modules, then the sequence of their $\mathfrak{m}$-completions

$$0 \to \hat{M}' \to \hat{M} \to \hat{M}'' \to 0$$

is also exact. In other words, the functor $M \mapsto \hat{M}$ is exact on the category of finitely generated modules over a noetherian ring A.

Proof. The $\mathfrak{m}$-topology of M induces the $\mathfrak{m}$-topology on M' by the Artin-Rees lemma (1.4., c)). The statement is then a consequence of Propositions 2.10 and 2.17. □

Theorem 4.6. *If M is a finitely generated module over a noetherian ring A, $\mathfrak{m}$ an ideal of A, $\hat{A}$ and $\hat{M}$ the $\mathfrak{m}$-completions of A*

and M respectively, then the canonical mapping $\hat{A}\otimes_A M \to \hat{M}$ *is an isomorphism.*

Proof. As M is finitely generated we have an exact sequence

$$0 \to N \to L \to M \to 0$$

where L is a finitely generated free module and N is also finitely generated since A is noetherian. The following diagram

$$\begin{array}{ccccccc} \hat{A}\otimes_A N & \longrightarrow & \hat{A}\otimes_A L & \longrightarrow & \hat{A}\otimes_A M & \longrightarrow & 0 \\ \downarrow{\scriptstyle \chi} & & \downarrow{\scriptstyle \psi} & & \downarrow{\scriptstyle \varphi} & & \\ \hat{N} & \longrightarrow & \hat{L} & \longrightarrow & \hat{M} & \longrightarrow & 0 \end{array}$$

is commutative and its rows are also exact by theorem 4.5. Furthemore ψ is an isomorphism (by lemma 4.1.) and φ, χ are onto by theorem 4.2. Then it follows by diagram chasing that φ is injective and so bijective. ☐

The results obtained so far allow us to prove a flatness theorem for $\mathfrak{m}$-completions. We recall first some definitions and results about flat and faithfully flat modules.

An A-module E is *flat* if for any exact sequence

$$(1) \qquad M' \to M \to M''$$

of A-modules, the corresponding sequence

$$(2) \qquad E\otimes M' \to E\otimes M \to E\otimes M''$$

is exact. If the converse is also true, we say that E is *faithfully flat*. In other words E is flat if the fuctor $M \mapsto E\otimes_A M$ is exact on the category of A-modules, and is faithfully flat if the above functor is exact and faithful.

We recall two useful propositions which characterize flat and faithfully flat modules respectively:

Proposition 4.7. *Let A be a ring and E an A-module. Then the following conditions are equivalent*

(i) *E is flat;*

(ii) *for any exact sequence* $0\to M' \to M \to M'' \to 0$ *of A-modules, the corresponding sequence* $0 \to E\otimes M' \to E\otimes M \to E\otimes M'' \to 0$ *is exact;*

(iii) *for any injective homomorphism* $f: M' \to M$ *of finitely generated A-modules, the homomorphism* $1 \otimes f: E \otimes M' \to E \otimes M$ *is injective*

Proof. See [3] p. 29, prop. 2.19. □

Proposition 4.8. *Let A be a ring and E an A-module. Then the following conditions are equivalent:*

(i) *E is faithfully flat;*

(ii) *E is flat, and for any A-module N the relation* $E \otimes N = 0$ *implies* $N = 0$*;*

(iii) *E is flat, and for any homomorphism* $f: N' \to N$ *of A-modules, the relation* $1_E \otimes f = 0$ *implies* $f = 0$.

(iv) *E is flat, and for any maximal ideal* $\mathfrak{m}$ *of A one has* $\mathfrak{m}E \neq E$.

Proof. See [5], p. 44, prop. 1. □

Now we can give the flatness theorem.

Theorem 4.9. *Let A be a noetherian* $\mathfrak{m}$*-adic ring and* $\hat{A}$ *its completion. Then:*

(i) $\hat{A}$ *is a flat A-module.*

(ii) $\hat{A}$ *is faithfully flat if and only if A is a Zariski ring.*

Proof. In order to prove (i) we use condition (iii) of Proposition 4.7. Let $f: M' \to M$ be an injective homomorphism of finitely generated A-modules. Then the homomorphism $\hat{f}: \hat{M}' \to \hat{M}$ is injective by Theorem 4.5, and (i) follows immediately by Theorem 4.6.

By Corollary 2.19 we have that A is a Zariski ring (i.e. $\mathfrak{m} \subset \operatorname{rad} A$) if and only if $\mathfrak{n}\hat{A} \neq \hat{A}$ for any maximal ideal $\mathfrak{n}$ of A. Thus (ii) follows by (i) and condition (iv) of Proposition 4.8. □

Remark. The above theorem is false if A is not noetherian. A counterexample can be found in [6], p. 120, ex. 14, b), where a local subring A of $k[[X, Y]] = B$ is given such that $\hat{A} = B$, but B is not A-flat.

Another counterexample is the following. Let $T, Y, X_0, X_1, \ldots$ be indeterminates over a field k,

$$B = k[Y, X_0, X_1, \ldots]/(YX_0, \ldots, X_i - YX_{i+1}, \ldots), \qquad A = B[T]$$

and $\mathfrak{m} = (T)A$. Then $\hat{A} = B[[T]]$ is not flat over A, by condition (iii) of Proposition 4.7, since $y - T$ is not a zero divisor in A, while in $\hat{A}$ one has $(y-T)\sum x_i T^i=0$; (here x_i and y denote the classes of X_i and Y in B).

Corollary 4.10. *Let A be a noetherian $\mathfrak{m}$-adic ring, and $B=(A,\mathfrak{m})\{X_1,\ldots,X_n\}$ the ring of restricted power series. Then:*

(i) *B is flat over $A[X_1,\ldots,X_n]$.*

(ii) *B is faithfully flat over A.*

Proof. Since B is the completion of the polynomial ring $A[X_1,\ldots,X_n]$ with respect to the $\mathfrak{m}(X_1,\ldots,X_n)$-topology (Prop. 3.6), (i) follows immediately by Theorem 4.9.

Since $A[X_1,\ldots,X_n]$ is a free (and hence flat) A-module, it follows easily from (i) that B is flat over A. Moreover if $\mathfrak{n}$ is a maximal ideal of A, one has $\mathfrak{n}B\neq B$, and (ii) follows from condition (iv) of Proposition 4.8. ▯

Corollary 4.11. *Let A be a noetherian ring and $B=A[[X_1,\ldots,X_n]]$. Then*

(i) *B is flat over $A[X_1,\ldots,X_n]$.*

(ii) *B is faithfully flat over A.*

Proof. Apply Corollary 4.10 with $\mathfrak{m}=A$. ▯

Theorem 4.9 has many other consequences (see e.g. [5], Chapter I and [6], § 3, n. 4.5).

§ 5
Noetherian properties of $\mathfrak{m}$-adic completions

The aim of this paragraph is to show that the noetherian property is preserved by $\mathfrak{m}$-adic completions. Using some facts stated in § 3, it is now possible to give an immediate proof of this result.

Theorem 5.1. *If A is a noetherian ring and $\mathfrak{m}$ an ideal of A, then the $\mathfrak{m}$-completion $\hat{A}$ of A is also noetherian.*

Proof. Let $\mathfrak{m}$ be generated by the elements $a_1, \ldots, a_n$, so that the $\mathfrak{m}$-completion $\hat{A}$ of A is isomorphic to the ring $A[[X_1, \ldots, X_n]]/(X_1 - a_1, \ldots, X_n - a_n)$. The ring $A[[X_1, \ldots, X_n]]$ is noetherian by Theorem 3.2, hence the result. □

Remark. It is clear that the above proof is correct only if we choose for theorem 3.2 a proof which is independent of Theorem 5.1, for example one of the direct proofs mentioned in § 3. On the other hand we will give an alternate proof of Theorem 5.1 which does not depend on Theorem 3.2; following this procedure we may get Theorem 3.2 as an immediate corollary of Theorem 5.1. The crucial point of this second proof of Theorem 5.1 is to be found in Theorem 5.3, where sufficient conditions for some $\mathfrak{m}$-complete rings to be noetherian are given.

We shall not give the proofs of all results, since the sequence of the following facts is essentialy the same as in [6], Ch. III, § 2, 3 or [3], Ch. 10, to which the reader is referred for all details.

The following lemma is used in the proof of Theorem 5.3 and has also an interest in itself, as it shows how some properties of the associated graded rings are reflected in the completions.

Lemma 5.2. *Let A, B be filtered groups and $\varphi: A \to B$ a homomorphism such that $\varphi(A_n) \subset B_n$. Let $Gr(\varphi): Gr(A) \to Gr(B)$ and $\hat{\varphi}: \hat{A} \to \hat{B}$ be the canonical homomorphisms induced by φ on the associated graded groups and completed groups respectively.*

Then:

(i) $Gr(\varphi)$ *injective* $\Rightarrow \hat{\varphi}$ *injective,*

(ii) $Gr(\varphi)$ *surjective* $\Rightarrow \hat{\varphi}$ *surjective.*

Proof. See, for example, [3], Prop. 10.2. ▯

We now observe that if M is an A-module and if (M_n) is an $\mathfrak{m}$-filtration of A, then $Gr(M) = \oplus M_n/M_{n+1}$ is a graded $Gr(A)$-module where $Gr(A)$ is the graded ring associated to the $\mathfrak{m}$-filtration of A.

Theorem 5.3. *Let A be a ring, $\mathfrak{m}$ un ideal of A such that A is complete for the $\mathfrak{m}$-topology, M an A-module, (M_n) an $\mathfrak{m}$-filtration of M inducing on M a Hausdorff topology. Then:*

(i) $Gr(M)$ *finitely generated* $\Rightarrow M$ *finitely generated,*

(ii) $Gr(M)$ *noetherian* $\Rightarrow M$ *noetherian.*

Proof. See, for example, [3] Prop. 10.24 and 10.25. ▯

Corollary 5.4. *If A is complete (hence Hausdorff) for the $\mathfrak{m}$-topology, then: $Gr(A)$ noetherian $\Rightarrow A$ noetherian.*

We now give the second proof of Theorem 5.1.

Corollary 5.5. (= Theorem 5.1). *A noetherian $\Rightarrow \hat{A}$ noetherian.*

Proof. It follows immediately from Corollary 5.4 and the isomorphism $Gr(A) \cong Gr(\hat{A})$ (see 2.15). ▯

It is now clear that in this way Theorem 3.2 can be considered a particular case of Theorem 5.1: in fact $A[[X_1, \ldots, X_n]]$ is the $(X_1, \ldots, X_n)$-completion of $A[X_1, \ldots, X_n]$ which is noetherian (by the Hilbert basis theorem) when A is noetherian. It is also evident that Theorem 3.2 follows immediately by Corollary 5.4 as well as by the next result.

Corollary 5.6. *Let A be a complete (Hausdorff) ring with respect to its $\mathfrak{m}$-topology; suppose that $A/\mathfrak{m}$ is noetherian and $\mathfrak{m}/\mathfrak{m}^2$ is a finitely generated $(A/\mathfrak{m})$-module. Then A is noetherian.*

Proof. The ring $Gr(A)$ is noetherian, since it is an $(A/\mathfrak{m})$-graded algebra generated by $\mathfrak{m}/\mathfrak{m}^2$; whence the result by Corollary 5.4. ▯

Remark 1. The first statement of Theorem 5.3 may be considered a partial converse of the following proposition, whose proof is easy (see [3], Prop. 10.22, (iii)).

Proposition 5.7. *If M is a finitely generated A-module and (M_n) is a stable $\mathfrak{m}$-filtration of M, then $Gr(M)$ is finitely generated over $Gr(A)$.*

Remark 2. The converse of Theorem 5.1 is not true. In other words there exist non noetherian rings A whose $\mathfrak{m}$-completions $\hat{A}$ are noetherian for some ideals $\mathfrak{m}$ of A. Thus noetherianness is an ascending but not a descending property for $\mathfrak{m}$-completions. A counterexample to this descending property is given in [6], Ex. 14, p. 119: there exists a local non noetherian subring B of $k[[X,Y]]$ (k is a field) whose completion (with respect to the maximal ideal) is isomorphic to $k[[X,Y]]$.

We notice that this counterexample was already mentioned in § 4.

Proposition 5.8. *Let $(A,\mathfrak{m})$ be a noetherian $\mathfrak{m}$-adic ring. Then the ring $B=(A,\mathfrak{m})\{X_1,\ldots,X_n\}$ of restricted power series is also noetherian.*

Proof. Since B is the $\mathfrak{m}(X_1,\ldots,X_n)$-completion of $A[X_1,\ldots,X_n]$ (prop. 3.6), the conclusion follows from Theorem 5.1.

§ 6
Some general criteria of ascent and descent for $\mathfrak{m}$-completions

In the last paragraph we have seen that the $\mathfrak{m}$-completion of a noetherian ring is a noetherian ring, while the converse is not true.

In the following sections we shall study the relations between A and $\hat{A}$ with respect to some other properties and we shall give sufficient and/or necessary conditions in order to have these properties verified by $\hat{A}$. More precisely we are interested in the following properties: dimension, regularity, "Cohen-Macaulay", "Gorenstein", integrity, unique factorization, property S_n, property R_n, reducedness, normality.

Sometimes A and $\hat{A}$ are independent with respect to some of these properties. We may consider, for example, the unique factorization property: A may be factorial while $\hat{A}$ is not factorial and conversely. Sometimes there is a dependence in only one sense, as for the noetherian property.

A ring homomorphism $f: A \to B$ is said to be *(faithfully) flat* if f induces on B the structure of a (faithfully) flat A-module.

Lemma 6.1. *Let* $f: A \to B$ *be a faithfully flat homomorphism,* $\mathfrak{n}$ *an ideal of* B, *and* $\mathfrak{m} = f^{-1}(\mathfrak{n})$. *Suppose that* $\mathfrak{n} = \mathfrak{m}B$ *and that* $A/\mathfrak{m} \cong B/\mathfrak{n}$. *Then the canonical ring homomorphism* $\hat{f}: (A, \mathfrak{m})\hat{} \to (B, \mathfrak{n})\hat{}$ (prop. 2.17) *is bijective.*

Proof: Since $\hat{A} = \varprojlim A/\mathfrak{m}^k$ and $\hat{B} = \varprojlim B/\mathfrak{n}^k$ (prop. 2.16), it is sufficient to show that the homomorphisms $f_k: A/\mathfrak{m}^k \to B/\mathfrak{n}^k$ induced by f are isomorphisms ($k = 1, 2, \ldots$).

Now, as $\mathfrak{m}B = \mathfrak{n}$ we get $\mathfrak{m}^k B = \mathfrak{n}^k$, and since f is faithfully flat, we have $f^{-1}(\mathfrak{n}^k) = \mathfrak{m}^k$ ([5] p. 51, prop. 9). This shows that

f_k is injective. Moreover, since $A/\mathfrak{m} = B/\mathfrak{m}B$, we have

$$B = f(A) + \mathfrak{m}B = f(A) + \mathfrak{m}^2 B = \cdots = f(A) + \mathfrak{m}^k B$$

whence f_k is onto. □

Proposition 6.2. *Let A be a noetherian $\mathfrak{m}$-ring, $\hat{A}$ its completion, $\mathfrak{N}$ a maximal ideal of $\hat{A}$ and $\mathfrak{n}$ its contraction to A,* (*hence $\mathfrak{n}$ is a maximal ideal containing $\mathfrak{m}$, by* 2.19). *Then:*

a) $\mathfrak{n}\hat{A}_{\mathfrak{N}} = \mathfrak{N}\hat{A}_{\mathfrak{N}}$ *and* $A/\mathfrak{n} = \hat{A}/\mathfrak{N}$.

b) $\hat{A}_{\mathfrak{N}}$ *is faithfully flat over* $A_{\mathfrak{n}}$.

c) *The completions of the local rings $\hat{A}_{\mathfrak{N}}$ and $A_{\mathfrak{n}}$ are canonically isomorphic.*

Proof. By Corollary 2.20 we have easily a). Moreover $\hat{A}$ is flat over A, whence $\hat{A}_{\mathfrak{N}}$ is flat over $A_{\mathfrak{n}}$, ([5], p. 116, prop. 15). But $\mathfrak{n}\hat{A}_{\mathfrak{N}} \neq \hat{A}_{\mathfrak{N}}$ and thus $\hat{A}_{\mathfrak{N}}$ is faithfully flat. This proves b). Finally c) follows immediately from a), b) and Lemma 6.1. □

It is now convenient to introduce the following terminology.

Let $f: A \to B$ be a ring homomorphism and let P be any property of commutative rings. We say that P *ascends* by f (or from A to B) if the following fact is true: A has property $P \Rightarrow B$ has property P. If the converse implication is true, we say that P *descends* from B to A.

A property P of commutative rings is said to be *local* if the following is true: A has $P \Leftrightarrow A_{\mathfrak{m}}$ has P for each maximal ideal $\mathfrak{m}$ of A. We say also that *A has the property P at the prime ideal* $\mathfrak{p}$ if $A_{\mathfrak{p}}$ has property P.

Proposition 6.3. *Let A be a noetherian $\mathfrak{m}$-adic ring, $\hat{A}$ its completion, $\mathfrak{N}$ a maximal ideal of $\hat{A}$ and $\mathfrak{n}$ its contraction to A. Let P be a local property which descends by faithful flatness. Then, if $\hat{A}_{\mathfrak{N}}$ has property P so does $A_{\mathfrak{n}}$. In particular if $\hat{A}$ has property P then A has property P at all maximal ideals containing $\mathfrak{m}$.*

Proof. By Proposition 6.2 $\hat{A}_{\mathfrak{N}}$ is faithfully flat over $A_{\mathfrak{n}}$, and the first statement follows. The last assertion follows then from Corollary 2.19. □

The above proposition is very useful since we shall always consider local properties which descend by faithful flatness.

Proposition 6.4. *Let A be a noetherian $\mathfrak{m}$-ring, and $\hat{A}$ its $\mathfrak{m}$-completion. Let P be any local property such that:*

a) *a local ring has P if and only if its completion does.*

Then the following conditions are equivalent:

(i) *$\hat{A}$ has P;*

(ii) *A has P at any maximal ideal containing $\mathfrak{m}$.*

If moreover $\mathfrak{m} \subset \operatorname{rad} A$, the above conditions are equivalent to:

(iii) *A has P.*

Proof. It follows easily from Proposition 6.2 and Corollary 2.19. □

Remark. Using the above Proposition we shall show that properties such as regularity, Cohen-Macaulay and Gorenstein ascend from A to $\hat{A}$, and descend if $(A, \mathfrak{m})$ is a Zariski ring. We shall see, in fact, that condition a) of Proposition 6.4 holds if P is "regularity", etc.

Proposition 6.5. *Let A be a noetherian $\mathfrak{m}$-ring and $\hat{A}$ its completion. Let P be any local property such that:*

a) *P descends by faithful flatness;*

b) *for any maximal ideal $\mathfrak{n}$ of A containing $\mathfrak{m}$, the local ring $(A_{\mathfrak{n}})^{\wedge}$ has P.*

Then $\hat{A}$ has P.

Proof. Let $\mathfrak{N}$ be a maximal ideal of $\hat{A}$, and let $\mathfrak{n}$ be the contraction of $\mathfrak{N}$ to A. Then the completions of $A_{\mathfrak{n}}$ and $\hat{A}_{\mathfrak{N}}$ are isomorphic by Proposition 6.2, (iii).

By condition b) $(A_{\mathfrak{n}})^{\wedge}$ has property P. But since $\hat{A}_{\mathfrak{N}}$ is noetherian, $(A_{\mathfrak{n}})^{\wedge}$ is faithfully flat over $\hat{A}_{\mathfrak{N}}$ (th. 4.9.); the conclusion follows by condition a). □

Remark 1. Condition a) of Proposition 6.5 is true for all local properties we are interested in. Thus if we want to see whether such a local property P ascends from A to $\hat{A}$, we can restrict our investigation to the ascent from A to $(A_{\mathfrak{n}})^{\wedge}$ as $\mathfrak{n}$ ranges over the set of all maximal ideals of A containing $\mathfrak{m}$. The ascent criteria given by Proposition 6.5 will be applied to properties such as S_n, R_n, reducedness, normality, as Proposition 6.4 is not useful in those cases. We shall also see that particularly simple sufficient conditions of ascent can be given using the concept of *formal fibers* introduced by Grothendieck.

Remark 2. We shall see that all the properties listed at the beginning of this section descend from $\hat{A}$ to A if A is a Zariski ring. For some of them (integrity and reducedness) this is trivial, since the canonical homomorphism $A \to \hat{A}$ is injective. Others descend by faithful flatness (regularity, Cohen-Macaulay, Gorenstein, R_n, S_n, normality), and thus the descent from $\hat{A}$ to A can follow from Theorem 4.9 (however direct proofs are sometime simpler).

Finally there are properties such as dimension and unique factorization, which do not descend by faithful flatness, but descend from $\hat{A}$ to A if A is a Zariski ring (as for dimension, we have in this case $\dim A = \dim \hat{A}$: see Theorem 7.5). We conclude by observing that the above properties may not descend if A is not a Zariski ring: it is not difficult to give counterexamples.

§ 7
Dimension of $\mathfrak{m}$-completions

We recall the classical definition of the dimension (Krull dimension) of a commutative ring A.

Definition. Let A be a (commutative) ring. The supremum of the lenghts of all strictly increasing chains of prime ideals in A is called the *dimension* of A and is noted $\dim A$. We have $\dim A \in \mathbb{N}$, or $\dim A = +\infty$.

In this section we shall always suppose that all rings are noetherian; nevertheless, this assumption does not imply that our rings have finite dimension. Anyhow a semilocal noetherian ring has always a finite dimension.

We also recall that the *height* $h(\mathfrak{p})$ of a prime ideal $\mathfrak{p}$ of a ring A is defined as the supremum of the lengths of all strictly ascending chains of prime ideals ending at $\mathfrak{p}$. If $\mathfrak{a}$ is any ideal, we define the height of $\mathfrak{a}$ in this way: $h(\mathfrak{a}) = \min h(\mathfrak{p})$ where $\mathfrak{p}$ ranges over the set of all minimal prime ideals of $\mathfrak{a}$.

We have evidently: $\dim A = \sup h(\mathfrak{m})$ where $\mathfrak{m}$ ranges over the set of all maximal ideals $\mathfrak{m}$ of A.

Finally we recall that a *composition series* of an A-module M is a maximal chain of strictly decreasing submodules of M. If A is an artinian ring (i.e. a noetherian ring of dimension 0) and M is a finitely generated A-module, then every composition series of M has the same length $l(M)$, called the *length* of M (see, for example [3], Propositions 6.5, 6.7, 6.8).

We now suppose that A is a local ring and $\mathfrak{m}$ is its maximal ideal (more generally we could consider a pair $(A, \mathfrak{m})$ where A is a semilocal ring and $\mathfrak{m}$ its radical). Then we have the following statement concerning the so-called Hilbert-Samuel characteristic polynomial.

Theorem 7.1. *Let A be a local ring, $\mathfrak{m}$ its maximal ideal, $\mathfrak{q}$ an $\mathfrak{m}$-primary ideal, M a finitely generated A-module. Then the length $l(M/\mathfrak{q}^n M)$ is finite for all n; moreover for all sufficiently large n this length is a polynomial $g(n)$ whose degree does not depend on $\mathfrak{q}$.*

Proof. See, for example, [3], Propositions 11.4 and 11.6. □

Theorem 7.2. (Dimension theorem). *Let A be a local ring, $\mathfrak{m}$ its maximal ideal, $\mathfrak{q}$ an $\mathfrak{m}$-primary ideal. Then the following three integers are equal:*

(i) $\dim A$;

(ii) *the degree of the polynomial $l(A/\mathfrak{q}^n)$ (for large n);*

(iii) *the least number of generators of an $\mathfrak{m}$-primary ideal.*

Proof. See, for example, [3], th. 11.14. □

Remark. Proofs of Theorems 7.1 and 7.2 in the semilocal case can be found, for example, in any of these books: [13], [29], and [41].

Definition. If $x_1, \ldots, x_d$ generate an $\mathfrak{m}$-primary ideal and $d = \dim A$, we call $x_1, \ldots, x_d$ a *system of parameters of A.*

It is obvious that the integer (iii) of 7.2 may be replaced by the following one: the cardinality of a system of parameters of A.

By (i) = (ii) of 7.2 and the isomorphism $A/\mathfrak{m}^n \cong \hat{A}/(\mathfrak{m}^n)\hat{}$ we get immediately the following proposition.

Proposition 7.3. *If A is a local ring, and $\hat{A}$ is its completion then* $\dim A = \dim \hat{A}$.

In order to give some properties of the dimension of an $\mathfrak{m}$-completion in the general case, i.e. in the case of a pair $(A, \mathfrak{m})$ where A is any ring and $\mathfrak{m}$ any ideal of A, the following lemma will be useful.

Lemma 7.4. *If S is a multiplicative system in the ring A, then* $\dim S^{-1}A \leqslant \dim A$. *Moreover, if $\mathfrak{p}$ is a prime ideal of A, we have* $\dim A_{\mathfrak{p}} = h(\mathfrak{p}) \leqslant \dim A$.

Proof. It follows immediately from the bijective correspondence between the prime ideals of $S^{-1}A$ and the prime ideals of A which don't meet S. □

Suppose now that $\mathfrak{m} \subset \operatorname{rad} A$ and let $\hat{A}$ be the $\mathfrak{m}$-completion of A. Then there is a bijective correspondence between the set of all maximal ideals $\mathfrak{N}$ of $\hat{A}$ and the set of all maximal ideals $\mathfrak{n} = \mathfrak{N} \cap A$ of A. (Cor. 2.19). We have $h(\mathfrak{N}) = \dim \hat{A}_{\mathfrak{N}}$ and $h(\mathfrak{n}) = \dim A_{\mathfrak{n}}$. The local rings $A_{\mathfrak{n}}$ and $\hat{A}_{\mathfrak{N}}$ have isomorphic completions by 6.2 and hence the same dimension by 7.3. Then it follows that $h(\mathfrak{N}) = h(\mathfrak{n})$ and hence $\dim A = \dim \hat{A}$.

If $\mathfrak{m} \not\subset \operatorname{rad} A$ the above conclusion is not necessarily true. In this case we have a bijection between the maximal ideals of $\hat{A}$ and the maximal ideals of A which contain $\mathfrak{m}$. Then it follows, by the above arguments, that $\dim \hat{A} = \sup h(\mathfrak{n})$ where $\mathfrak{n}$ ranges over the set of the maximal ideals of A containg $\mathfrak{m}$. Hence we have the inequality $\dim \hat{A} \leqslant \dim A$.

We may get the same conclusion, if we consider the zariskification $S^{-1}A$ of A where $S = 1 + \mathfrak{m}$. Then we have by 7.4 and the preceding result for Zariski rings: $\dim A \leqslant \dim S^{-1}A = \dim \hat{A}$.

We collect our results in the following statement.

Theorem 7.5. *Let $\hat{A}$ be the $\mathfrak{m}$-completion of A. Then* $\dim \hat{A} \leqslant \dim A$ *and equality holds if* $\mathfrak{m} \subset \operatorname{rad} A$.

Remark. If $\mathfrak{m}$ is a maximal ideal of A such that $h(\mathfrak{m}) < \dim A$, then the $\mathfrak{m}$-completion of A is a local ring $\hat{A}$ such that $\dim \hat{A} = h(\mathfrak{m}) < \dim A$ and the previous inequality is strict. The above situation is realized in the following example. Let $A = k[X, Y]/(YX, Y(Y-1)) = k[x, y]$ where k is a field; if $\mathfrak{m} = (x, y-1)$, then $\mathfrak{m}$ is a maximal ideal such that $h(\mathfrak{m}) = 0$, while $\dim A = 1$.

We conclude with the following thorem.

Theorem 7.6. *Let A be a ring, $X_1, \ldots, X_n$ indeterminates over A, $\mathfrak{m}$ an ideal of A. Then:*

a) $\dim A[X_1, \ldots, X_n] = \dim A + n$;

b) $\dim (A, \mathfrak{m})\{X_1, \ldots, X_n\} = \dim A + n$;

c) $\dim A[[X_1, \ldots, X_n]] = \dim A + n$.

Proof. As for a), see [29], 9.10. Then b) follows easily from a) and 7.5: all details can be found in [30], Th. 3. Finally c) is a particular case of b), when $\mathfrak{m} = A$.

§ 8
Regularity and global dimension of $\mathfrak{m}$-completions

In this paragraph we shall discuss ascent and descent of regularity for completions and, in particular, for rings of formal and restricted power series. *In this section all rings are noetherian.*

We begin with a theorem which characterizes regular local rings. We recall that if A is a local ring with maximal ideal $\mathfrak{m}$, we denote by $Gr(A)$ the graded ring $\oplus\mathfrak{m}^n/\mathfrak{m}^{n+1}$ associated with the $\mathfrak{m}$-topology of A.

Theorem 8.1. *Let A be a local ring of dimension d, $\mathfrak{m}$ its maximal ideal, and $k=A/\mathfrak{m}$. Then the following conditions are equivalent:*
(i) $Gr(A)\cong k[X_1,\dots,X_d]$, *where $X_1,\dots,X_d$ are indeterminates;*
(ii) $\dim_k \mathfrak{m}/\mathfrak{m}^2=d$;
(iii) $\mathfrak{m}$ *can be generated by d elements.*

Proof. See, for example, [3], Th. 11.22. □

Definition 8.2. A local ring A is said to be *regular* if it satisfies the equivalent conditions of Theorem 8.1. In general, a ring A is called *regular* if $A_\mathfrak{m}$ is a regular local ring for any maximal ideal $\mathfrak{m}$.

Examples. a) A field is a regular local ring of dimension 0.
b) if k is a field, and $X_1,\dots,X_d$ are indeterminates, the local ring $k[[X_1,\dots,X_d]]$ is regular.
c) A discrete valuation ring is a local regular ring of dimension 1. It follows that a Dedekind domain (in particular $\mathbb{Z}$) is regular.

Theorem 8.3. *Let A be an $\mathfrak{m}$-adic ring and $\hat{A}$ its completion. Then the following conditions are equivalent:*
(i) *$\hat{A}$ is regular;*
(ii) *$A_\mathfrak{n}$ is regular for any maximal ideal $\mathfrak{n}$ containing $\mathfrak{m}$. If moreover $\mathfrak{m}\subset \operatorname{rad} A$, the above conditions are equivalent to*
(iii) *A is regular.*

Proof. By Proposition 6.4 it is sufficient to show that a local ring B is regular if and only if $\hat{B}$ is regular. This follows immediately from condition (ii) of Theorem 8.1, since $Gr(B) \cong Gr(\hat{B})$ (prop. 2.15, c)). ▯

Regularity can be characterized by means of projective dimension of modules. We recall that if M is a module over a (not necessarily noetherian) ring A, the *projective dimension* of M ($p\dim_A M$) is the least integer n such that there is an exact sequence $0 \to P_n \to P_{n-1} \to \cdots \to P_0 \to M \to 0$, where $P_0, \ldots, P_n$ are projective A-modules (a module P is *projective* if there is a module Q such that $P \oplus Q$ is free). The *global dimension* of A ($gl\dim A$) is defined as the supremum of the projective dimensions of all A-modules.

We have the following characterization of a regular local ring (essentially it is a restatement, due to Serre, of the celebrated Hilbert Syzygies Theorem in the local case).

Theorem 8.4. *Let A be a (noetherian) local ring. Then the following conditions are equivalent:*

(i) *A is regular;*

(ii) $gl\dim A < \infty$;

(iii) $gl\dim A = \dim A$.

Proof. See, for example, [39], Chap. IV-D, p. IV-35. ▯

Corollary 8.5. *Let A be a ring. Then:*

(i) *If A is regular, we have $gl\dim A = \dim A$.*

(ii) *If $gl\dim A < \infty$, then A is regular.*

Proof. It is known that $gl\dim A$ is the supremum of the $gl\dim A_{\mathfrak{m}}$, as $\mathfrak{m}$ ranges over the set of all maximal ideals of A (see, e.g., [21], 17.2.10). The conclusion is then an easy consequence of Theorem 8.4. ▯

The following Proposition describes the behaviour of projective and global dimension under flat and faithfully flat homomorphisms.

Proposition 8.6. *Let $\varphi: A \to B$ be a flat homomorphism of (not necessarily noetherian) rings, and let M be an A-module. Then:*

(i) $p\dim_B(B \otimes_A M) \leqslant p\dim_A M$.

(ii) *If A is noetherian, M is finitely generated, and φ is faithfully flat then* $p\dim_B(B \otimes_A M) = p\dim_A M$,

(iii) *If A is noetherian and φ is faithfully flat, then* $gl\dim A \leqslant gl\dim B$.

Proof. The first two statements follow by [22], p. 137, Prop. 6.2.1. Moreover, by a theorem of Auslander ([39], p. IV-29, corollaire) we have that $gl\dim A$ is the supremum of the projective dimensions of all *finitely generated* A-modules. Then (iii) follows from (ii). □

By means of the above Proposition we can show that regularity descends by faithful flatness.

Proposition 8.7. *Let $\varphi: A \to B$ be a faithfully flat ring homomorphism. Then if B is regular so is A.*

Proof. Let $\mathfrak{m}$ be a maximal ideal of A. Since φ is faithfully flat, there is a maximal ideal $\mathfrak{M}$ of B such that $\varphi^{-1}(\mathfrak{M}) = \mathfrak{m}$ (see, e.g. [5], p. 54, Prop. 9). Moreover the local homomorphism $\psi: A_{\mathfrak{m}} \to B_{\mathfrak{M}}$ induced by φ is faithfully flat, and then, by the above Proposition, we have $gl\dim A_{\mathfrak{m}} \leqslant gl\dim B_{\mathfrak{m}}$. The conclusion follows then by Theorem 8.4. □

Proposition 8.8. *Let A be an $\mathfrak{m}$-ring, M a finitely generated A-module, $\hat{A}$ and $\hat{M}$ the $\mathfrak{m}$-completions of A and M respectively. Then:*

(i) $p\dim_{\hat{A}} \hat{M} \leqslant p\dim_A M$.

(ii) *If* $\mathfrak{m} \subset \operatorname{rad} A$, *then* $p\dim_{\hat{A}} \hat{M} = p\dim_A M$.

Proof. Since M is finitely generated, we have $\hat{M} \cong \hat{A} \otimes M$ (Th. 4.6), and the conclusion follows easily from Proposition 8.6 and the flatness theorem for $\mathfrak{m}$-completions (Th. 4.9).

We can apply the above results to the global dimensions of $\mathfrak{m}$-completions. We get a result similar to theorem 7.4 on Krull dimension.

Theorem 8.9. *Let A be an $\mathfrak{m}$-ring, and $\hat{A}$ its completion. Then we have $gl\dim \hat{A} \leqslant gl\dim A$, and equality holds if $\mathfrak{m} \subset \operatorname{rad} A$.*

Proof. If $gl\dim A = \infty$ the first assertion is clear, and the second follows easily by Proposition 8.8 (ii). Otherwise we have that A is regular (Cor. 8.5) and hence $\hat{A}$ is regular. Thus we have

$\dim A = gl\dim A$ and $\dim \hat{A} = gl\dim \hat{A}$ (Cor. 8.5). The conclusion follows then from Theorem 7.4. ▯

Now we apply the above results to the rings of formal and restricted power series.

Theorem 8.10. *Let A be an $\mathfrak{m}$-ring, $B = (A, \mathfrak{m})\{X_1, \ldots, X_n\}$. Then:*
(i) *A is regular if and only if B is regular.*
(ii) *$gl\dim B = gl\dim A + n$.*

Proof. It is known that if A is regular, the polynomial ring $C = A[X_1, \ldots, X_n]$ is regular (see, e.g., [35], Th. 2.1.). Moreover B is the $\mathfrak{m}(X_1, \ldots, X_n)$-completion of C (Prop. 3.6), and hence B is regular by Theorem 8.3. Conversely, since B is faithfully flat over A (Cor. 4.10), if B is regular, A is regular by Proposition 8.7. This proves (i).

To prove (ii) we recall that $\dim B = \dim A + n$ (Th. 7.6). If A is regular the conclusion follows immediately by (i) and Corollary 8.5. If on the contrary A is not regular, B is not regular either; thus $gl\dim A = gl\dim B = \infty$, and the conclusion is true in this case too. ▯

Corollary 8.11. *Let A be a ring. Then:*
(i) *A is regular if and only if $A[[X_1, \ldots, X_n]]$ is regular.*
(ii) *$gl\dim(A[[X_1, \ldots, X_n]]) = gl\dim A + n$.*

Proof. Apply the above theorem with $\mathfrak{m} = A$. ▯

Remark. In the proof of Theorem 8.10 we have seen that if A is regular the polynomial ring $A[X_1, \ldots, X_n]$ is regular. In particular if k is a field, the polynomial ring $k[X_1, \ldots, X_n] = A$ is regular, that is $A_{\mathfrak{m}}$ is a regular local ring for any maximal ideal $\mathfrak{m}$ of A. More generally it is known from algebraic geometry that if A is the coordinate ring of an algebraic variety V, the local ring of A at any prime corresponding to a nonsingular subvariety (i.e. a subvariety containing a simple point) is regular.

§ 9
"Cohen Macaulay" and "Gorenstein" properties for $\mathfrak{m}$-completions

In this section we want to discuss Cohen-Macaulay rings, and a special case of them, that is the Gorenstein rings. *All rings are assumed to be noetherian.*

The first ones arise in two "unmixedness theorems" due to Macaulay and Cohen respectively (see the examples below). We recall that an ideal $\mathfrak{a}$ of the ring A is *unmixed* if all the prime ideals associated with $\mathfrak{a}$ have the same height (which is then equal to the height of $\mathfrak{a}$). Furthermore an ideal $\mathfrak{a}$ is said to be of the *principal class* if it can be generated by $s=h(\mathfrak{a})$ elements.

Definition 9.1. *A ring A is Cohen-Macaulay (CM for short) if each ideal of the principal class of A is unmixed.*

Examples. a) The theorem of Macaulay mentioned before asserts that a polynomial ring $k[X_1, \ldots, X_n]$ over a field is a CM ring (see [41], p. 203, Th. 26).

b) A regular local ring is CM. This result is due to Cohen for equicharacteristic regular local rings (see, e.g. [41] p. 310, th. 29), but is true in general ([41], p. 397); and since, as we shall see, CM is a local property, every regular ring is CM.

c) A local domain of dimension 1 or an integrally closed local domain of dimension 2 is CM (see, e.g., [41], p. 397, or [15]).

The CM property is a local one, as our next theorem shows. In order to state it, we recall some definitions. Let A be a ring, and let $a_1, \ldots, a_n$ be elements of A such that $(a_1, \ldots, a_n) \neq A$. We say that $a_1, \ldots, a_n$ form an *A-sequence* if a_1 is not a zero-divisor and $(a_1, \ldots, a_{i-1}):(a_i)=(a_1, \ldots, a_{i-1})$ for any $i>1$. If $\mathfrak{a}$ is a proper ideal of A, the *grade* of $\mathfrak{a}$ (denoted by $gr(\mathfrak{a})$) is the maximal length of the A-sequences contained in $\mathfrak{a}$. One can show that, in general,

$gr(\mathfrak{a}) \leqslant h(\mathfrak{a})$([15], prop. 1.13). If A is a local ring with maximal ideal $\mathfrak{m}$, the grade of $\mathfrak{m}$ is often called the *depth* of A (depth A).

Theorem 9.2. *Let A be a ring. Then the following conditions are equivalent:*

(i) *A is CM;*

(ii) *any ideal of A generated by an A-sequence is unmixed;*

(iii) *$gr(\mathfrak{a})=h(\mathfrak{a})$ for any ideal $\mathfrak{a}$ of A;*

(iv) depth $A_{\mathfrak{p}} = \dim A_{\mathfrak{p}}$ *for any prime (resp. maximal) ideal $\mathfrak{p}$ of A;*

(v) *$A_{\mathfrak{p}}$ is CM for any prime (resp. maximal) ideal $\mathfrak{p}$ of A.*

If moreover A is a local ring, the above conditions are equivalent to:

(vi) *there exists a system of parameters of A which is an A-sequence;*

(vii) *every system of parameters of A is an A-sequence.*

Proof. It follows easily by [15], Theorems 3.1 and 4.1. ▯

Theorem 9.3. *Let A be an $\mathfrak{m}$-ring and $\hat{A}$ its completion. Then the following conditions are equivalent:*

(i) *$\hat{A}$ is CM;*

(ii) *$A_{\mathfrak{n}}$ is CM for any maximal ideal $\mathfrak{n}$ of A containing $\mathfrak{m}$.*

If moreover $\mathfrak{m} \subset \operatorname{rad} A$, the above conditions are equivalent to:

(iii) *A is CM.*

Proof. By Theorem 9.2 CM is a local property; thus, by Proposition 6.4, it is sufficient to show that a local ring is CM if and only if its completion is. This follows by [41], p. 400, cor. 6. ▯

Theorem 9.4. *Let A be a ring, $\mathfrak{m}$ an ideal of A and $X_1, \dots, X_n$ n indeterminates over A. Then the following conditions are equivalent:*

(i) *A is CM;*

(ii) *$A[X_1, \dots, X_n]$ is CM;*

(iii) *$B=(A,\mathfrak{m})\{X_1, \dots, X_n\}$ is CM;*

(iv) *$A[[X_1, \dots, X_n]]$ is CM.*

Proof. The implication (i)⇒(ii) follows by [29], Theorem 25.10, while (ii)⇒(iii) by Theorem 9.3 and Proposition 3.6. Furthermore, since $A \cong B/(X_1, \dots, X_n)$ and $X_1, \dots, X_n$ form a B-sequence, if B is CM so is A (see, e.g., [15], prop. 5.1). Thus (iii)⇒(i), and the first three conditions are equivalent. Finally the equivalence (i)⇔(iv) is a special case of (i)⇔(iii), when $\mathfrak{m}=A$. □

A special class of CM ring is given by "Gorenstein" rings. They arise in algebraic geometry (see, e.g., [4]), but here we restrict ourselves to an ideal-theoretic definition.

Definition 9.5. A CM local ring is said to be *Gorenstein* if there is a system of parameters of A which generates an *irreducibile* ideal (i.e. an ideal which is not an intersection of two properly larger ideals). A ring A is *Gorenstein* if $A_{\mathfrak{m}}$ is Gorenstein for any maximal ideal $\mathfrak{m}$ of A.

Gorenstein rings can be characterised by the following theorem.

Theorem 9.6. *Let A be a ring. Then the following conditions are equivalent*

(i) *A is Gorenstein;*

(ii) *$A_{\mathfrak{p}}$ is Gorenstein for any prime ideal $\mathfrak{p}$ of A;*

(iii) *each ideal of the principal class of A is unmixed, and its primary components are irreducibile;*

(iv) *each ideal generated by an A-sequence is unmixed and its primary components are irreducibile.*

If moreover A is local, the above conditions are equivalent to:

(v) *every system of parameters of A generates an irreducible ideal.*

Proof. See [17], th. 6.1. and 6.3, or [4]. □

Examples. a) A regular local ring is Gorenstein, as follows easily from the definitions. The converse is not true in general (see e.g. [17], § 6).

b) More generally any "complete intersection" is Gorenstein (a complete intersection is a ring of the form $A/(a_1, \ldots, a_n)$ where A is regular and $a_1, \ldots, a_n$ form an A-sequence), see [4] or [17]. There are, however, Gorenstein rings which are not of this type.

c) A Gorenstein ring is CM, but not conversely (for counter-examples see [17], § 6).

d) By a therorem of P. Murthy ([28]) a CM factorial domain (see next § 11), which is also a quotient of a regular local ring, is Gorenstein.

Theorem 9.7. *Let A be an $\mathfrak{m}$-ring and $\hat{A}$ its completion. Then the following conditions are equivalent:*

(i) *$\hat{A}$ Gorenstein;*

(ii) *$A_{\mathfrak{n}}$ is Gorenstein for any maximal ideal $\mathfrak{n}$ of A containing $\mathfrak{m}$.*

If moreover $\mathfrak{m} \subset \operatorname{rad} A$, the above conditions are equivalent to:

(iii) *A is Gorenstein.*

Proof. Since Gorenstein is a local property, we can apply Proposition 6.4; that is it is sufficient to show that a local ring is Gorenstein if and only if its completion is. But this follows easily from the definition (details are given in [27]). □

Theorem 9.8. *Let A be a ring, $\mathfrak{m}$ an ideal of A, and $X_1, \dots, X_n$ n indeterminates. Then the following conditions are equivalent:*

(i) *A is Gorenstein;*

(ii) *$A[X_1, \dots, X_n]$ is Gorenstein;*

(iii) *$(A, \mathfrak{m})\{X_1, \dots, X_n\} = B$ is Gorenstein;*

(iv) *$A[[X_1, \dots, X_n]]$ is Gorenstein.*

Proof. We have that (i)⇒(ii) by [27], (Teor. 1 and Cor. 2), and (ii)⇒(iii) by theorem 8.13 and Proposition 3.6. Moreover since $X_1, \dots, X_n$ form a B-sequence, and $A \cong B/(X_1, \dots, X_n)$, if B is Gorenstein so is A ([17], Cor. 6.6). Thus (iii)⇒(i), and the first three properties are equivalent. The equivalence (i)⇔(iv) is a particular case of (i)⇔(iii), for $\mathfrak{m} = A$. □

Further properties of the rings considered in this section can be found in [24].

§ 10
Integrity of $\mathfrak{m}$-completions

In this section we show (Th. 10.6) that with suitable conditions on the rings A and $A/\mathfrak{m}$, the $\mathfrak{m}$-completion of A is a domain. *All rings are assumed here to be noetherian.*

Definition 10.1. A local ring A is said to be *analytically irreducible* if its completion $\hat{A}$ is a domain. In this case A itself must be a domain, since the canonical homomorphism $A \to \hat{A}$ is injective.

A ring A is *analytically irreducible at a prime ideal* $\mathfrak{p}$ if the local ring $A_\mathfrak{p}$ is analytically irreducible.

We observe that there are local domains which are not analytically irreducible. For example, if we consider the ring $A = \mathbb{C}[X, Y]_{(X,Y)}/(XY + X^3 + Y^3)$, then A is a domain. However, one can show that $\hat{A} \cong \mathbb{C}[[X, Y]]/(XY)$, which is not a domain.

Proposition 10.2. *Let A be a ring, $\mathfrak{m}$ an ideal of A and $\hat{A}$ the $\mathfrak{m}$-completion of A. Suppose further that A is analytically irreducible at any maximal ideal containing $\mathfrak{m}$. Then $\hat{A}$ is locally a domain.*

Proof. It follows immediately from Proposition 6.2.

Now we want to show that with a further assumption on $A/\mathfrak{m}$ the ring $\hat{A}$ is a domain. For this we need several lemmas.

Lemma 10.3. *Let A be a ring. Then the following conditions are equivalent:*

a) *A is locally a domain;*

b) *there are domains $A_1, \dots, A_n$ such that $A \cong A_1 \times \cdots \times A_n$.*

Proof. Suppose a) holds. Let $\mathfrak{p}_1, \dots, \mathfrak{p}_n$ be the minimal primes of A and put $A_i = A/\mathfrak{p}_i$ $(i = 1, \dots, n)$. Let $\varphi: A \to \prod A_i = B$ be the canonical homomorphism. We shall show that φ is bijective.

In order to do this, it is sufficient to prove that for any maximal ideal $\mathfrak{m}$ of A the induced homomorphism

$$\varphi_{\mathfrak{m}}: A_{\mathfrak{m}} \to B \otimes_A A_{\mathfrak{m}}$$

is bijective ([5], p. 111, th. 1). Now, since $A_{\mathfrak{m}}$ is a domain, there is a unique i such that $\mathfrak{p}_i A_{\mathfrak{m}} = (0)$ and $\mathfrak{p}_j A_{\mathfrak{m}} = A_{\mathfrak{m}}$ for $i \neq j$. This implies $B \otimes_A A_{\mathfrak{m}} \cong A_{\mathfrak{m}}$. Thus $\varphi_{\mathfrak{m}}$ is an isomorphism and a)⇒b). The converse is clear. ▯

We recall that if A is a ring (not necessarily noetherian), the set of all prime ideals is a topological space, called the *spectrum* of A and written spec A (see [5], p. 124).

Lemma 10.4. *Let A be a (not necessarily noetherian) ring. Then* spec A *is connected if and only if A contains no idempotents $\neq 0,1$.*

Proof. See [5], p. 132, cor. 2. ▯

Corollary 10.5. *Let A be a ring. Then the following conditions are equivalent:*

(i) *A is a domain;*

(ii) *A is a locally a domain, and contains no idempotents $\neq 0, 1$;*

(iii) *A is locally a domain, and* spec A *is connected.*

Proof. It is clear that (i)⇒(ii); the converse is true by lemma 10.3. The equivalence of (ii) and (iii) follows by Lemma 10.4, and the proof is complete. ▯

In order to apply the above corollary to completions we have to know something about the idempotents of $\hat{A}$. These are in 1-1 correspondence with the idempotents of $A/\mathfrak{m}$, as the following lemma shows.

Lemma 10.6. *Let A be an $\mathfrak{m}$-complete ring, a an element of A such that $\bar{a} \in A/\mathfrak{m}$ is idempotent. Then there is a unique idempotent $e \in A$ such that $\bar{e} = \bar{a}$.*

Proof. We must show that there is a unique $x \in \mathfrak{m}$ such that $e = a + x$ is idempotent. Such x is clearly a solution of the equation $f(X) = 0$ where

$$f(X) = X^2 + (2a-1)X + a^2 - a.$$

Since $a^2 - a \in \mathfrak{m}$, reducing modulo $\mathfrak{m}$ we have

$$\bar{f}(X) = X(X + 2\bar{a} - 1).$$

Now $2\bar{a}-1$ is a unit, since $(2\bar{a}-1)^2=4\bar{a}^2-4\bar{a}+1=1$, and thus the polynomials X and $X+2\bar{a}-1$ of $(A/\mathfrak{m})[X]$ are coprime. Thus we can apply Hensel's lemma (th. 3.7) to get a unique couple $g(X)$, $h(X)$ of monic polynomials such that:

$$f(X)=g(X)h(X),$$
$$\bar{g}(X)=X, \quad \bar{h}(X)=X+2\bar{a}-1.$$

Then we have $g(X)=X-x$, $x\in\mathfrak{m}$ and this x is the unique solution belonging to $\mathfrak{m}$ of the equation $f(X)=0$. This completes the proof. □

Corollary 10.7. *Let A be an $\mathfrak{m}$-ring and $\hat{A}$ its completion. Then* $\operatorname{spec}\hat{A}$ *is connected if and only if* $\operatorname{spec}(A/\mathfrak{m})$ *is connected.*

Proof. Since $\hat{A}$ is $\mathfrak{m}$-complete and $A/\mathfrak{m}\cong\hat{A}/\hat{\mathfrak{m}}$, the conclusion follows by Lemmas 10.4 and 10.6. □

Proposition 10.8. *Let A be an $\mathfrak{m}$-ring and $\hat{A}$ its completion. Then the following conditions are equivalent:*

(i) *$\hat{A}$ is a domain;*
(ii) *$\hat{A}$ is locally a domain, and $A/\mathfrak{m}$ has no idempotents $\neq 0, 1$;*
(iii) *$\hat{A}$ is locally a domain, and* $\operatorname{spec}(A/\mathfrak{m})$ *is connected;*
(iv) *$\hat{A}$ is locally a domain, and* $\operatorname{spec}\hat{A}$ *is connected.*

Proof. By Lemma 10.6 and Corollary 10.5 it follows easily that (i)⇔(ii), while (ii)⇔(iii) by Lemma 10.4, and (iii)⇔(iv) by Corollary 10.7. □

Corollary 10.9. *Let A be an $\mathfrak{m}$-ring and $\hat{A}$ its completion. Suppose that A is analytically irreducible at any maximal ideal containing $\mathfrak{m}$. Then the following conditions are equivalent:*

(i) *$\hat{A}$ is a domain;*
(ii) *$A/\mathfrak{m}$ has no idempotents $\neq 0,1$;*
(iii) $\operatorname{spec}(A/\mathfrak{m})$ *is connected;*
(iv) $\operatorname{spec}\hat{A}$ *is connected.*

Proof. By Proposition 10.2 $\hat{A}$ is locally a domain. The conclusion follows by Proposition 10.8. □

Now we give an application of the above theorem. We state first an easy lemma.

Lemma 10.10. *Let A be a Zariski ring, and $\hat{A}$ its completion. Then if $\hat{A}$ is a domain so is A.*

Proof. By Theorem 1.5 A is Hausdorff, whence the canonical homomorphism $A \to \hat{A}$ is injective. The conclusion follows. $\square$

Corollary 10.11. *Let A be a ring, $\mathfrak{m}$ an ideal of A, and $\hat{A}$ the $\mathfrak{m}$-completion of A. Then the following conditions are equivalent:*

a) *$\hat{A}$ is a regular domain;*

b) *$A_{\mathfrak{n}}$ is regular for any maximal ideal $\mathfrak{n}$ containing $\mathfrak{m}$, and* $\operatorname{spec}(A/\mathfrak{m})$ *is connected.*

If moreover $(A, \mathfrak{m})$ is a Zariski ring, the above conditions are equivalent to:

c) $\operatorname{spec}(A/\mathfrak{m})$ *is connected, and A is a regular domain.*

Proof. By Theorem 8.3, $\hat{A}$ is regular if and only if A is regular for any maximal ideal $\mathfrak{n}$ containing $\mathfrak{m}$. Since a regular local ring is a domain, the equivalence of a) and b) follows easily by Proposition 10.8. Moreover if $(A, \mathfrak{m})$ is a Zariski ring, it is easy to see, with the help of Lemma 10.7, that a) and b) imply c). Moreover it is clear that c)$\Rightarrow$b), and the proof is complete. $\square$

§ 11
Unique factorization of $\mathfrak{m}$-completions

A) Definition and first properties of factorial rings

Let A be an integral domain. We say that A is *factorial* (or a Unique Factorization Domain) if every element $a \in A$, $a \neq 0$ and non-unit, has an essentially unique decomposition in irreducible factors. Here "essentially" means "up to unit factors and permutations of the factors".

Proposition 11.1. *An integral domain A is factorial if and only if the following are true:*

a) *Every non zero element of A is a product of irreducible factors.*

b) *Every irreducible element of A is prime* (i.e. *it generates a prime ideal*).

Proof. As condition a) gives the existence of a decomposition, we have only to verify that condition b) is equivalent to the uniqueness of the decomposition: this is a very well known fact. □

Remark. Condition a) of 11.1 is verified if A is noetherian.

Theorem 11.2. *Let A be a noetherian integral domain. Then the following are equivalent:*

(i) *A is factorial;*

(ii) *every prime ideal in A of height* 1 *is principal;*

(iii) *the intersection of two principal ideals of A is principal.*

Proof. See, for example [36], Ch. I, § 2, and Ch. III, Th. 6. □

The factorial property localizes, as is shown by the following statement.

Proposition 11.3. *Let S be a multiplicative system in A. Then A factorial $\Rightarrow S^{-1}A$ factorial.*

Proof. See [36], p. 29, Th. 4. ▯

The converse of 11.3 is not true in general. We know only the following partial converse, due to Nagata, if very strict conditions are given on S.

Theorem 11.4. *Let A be a noetherian ring, S a multiplicative system of A generated by prime elements and suppose that $S^{-1}A$ is factorial. Then A is factorial.*

Proof. See [36], p. 31, Th. 5. ▯

We can also see that factoriality is not a local property (see next Example 3). Hence the following definition will be useful.

Definition. A ring A is said to be *locally factorial* if $A_{\mathfrak{m}}$ is factorial for all maximal ideals $\mathfrak{m}$ of A.

Theorem 11.5. *Let A be a noetherian integral domain. Then the following are equivalent:*

(i) *A is locally factorial;*

(ii) *every prime ideal in A of height* 1 *is an invertible ideal;*

(iii) *the intersection of two non zero principal ideals of A is an invertible ideal.*

Proof. See [16], Th. 3.3. ▯

We denote by $\operatorname{Pic} A$ (Picard group) the class group of rank 1 projective A-modules. If A is a noetherian ring or an integral domain, then $\operatorname{Pic} A$ is isomorphic to the class group of fractionary invertible ideals (see [5], p. 151, Cor. 2 and Remarque).

Proposition 11.6. *Let A be a noetherian ring. Then the following are equivalent:*

(i) $\operatorname{Pic} A = 0$;

(ii) *every invertible ideal of A is free.*

Suppose, in addition, that $\operatorname{spec} A$ *is connected. Then* (i) *and* (ii) *are equivalent to:*

(iii) *every projective ideal of A is free.*

Proof. See [16], Prop. 1.7. ▯

Proposition 11.7. *If A is a noetherian factorial domain, then* $\operatorname{Pic} A = 0$.

Proof. See [36], p. 89, lemme 10, or [16], Cor. 2.5. □

From the preceding properties we get the following criteria for factoriality which we will often apply.

Theorem 11.8. *Let A be a noetherian integral domain. Then A is factorial if and only if the following conditions are verified:*
a) $\operatorname{Pic} A = 0$.
b) *A is locally factorial.*

Proof. If A is factorial, then a) and b) are true by 11.3 and 11.7. Conversely, if A is locally factorial and $\operatorname{Pic} A = 0$, then every prime ideal of height 1 of A is free (that is principal) by 11.5 and 11.6; hence A is factorial by 11.2. □

The following theorem, due to Auslander and Buchsbaum, is well known.

Theorem 11.9. *A regular local ring is factorial.*

Proof. See [41], Appendix 7 or [29], 28.7 or [36], p. 88, Th. 11. □

Examples.

1. Any field and the ring $\mathbb{Z}$ are unique factorization domains.
2. If A is a factorial ring and X is an indeterminate over A, then $A[X]$ is also factorial (Gauss theorem). In particular the ring $k[X_1, \ldots, X_n]$ in n indeterminates over a field is factorial.
3. If A is a non principal Dedekind domain, then we have $\operatorname{Pic} A \neq 0$. Therefore, by 11.8, A is not factorial. However A is a regular ring, hence locally factorial by virtue of Th. 11.9 and the fact that regularity is a local property. This example shows that factoriality is not a local property.
4. The coordinate ring of a circle over the field $\mathbb{R}$ (resp. $\mathbb{C}$) is a Dedekind *non factorial* domain (resp. *factorial*) (see [36], p. 30, Examples). This fact can be considered as an application of the following geometric "interpretation" of factoriality: the coordinate ring of an irreducible variety V is factorial if all irreducible subvarieties of V of codimension 1 are complete intersections (of V with a hypersurface) (see [36], § 4, p. 38).

From now on all rings are supposed to be noetherian.

B) The descent of factoriality from $\hat{A}$ to A for Zariski rings. A criterion for the factoriality of $\hat{A}$

Factoriality descends from $(A, \mathfrak{m})\hat{}$ to A if $(A, \mathfrak{m})$ is a Zariski ring. In other words we have the following result (due to Mori).

Theorem 11.10. *Let $(A, \mathfrak{m})$ be a Zariski ring such the $\mathfrak{m}$-completion $\hat{A}$ of A is a factorial domain. Then A is also factorial.*

Proof. See [36], p. 55, Th. 9. ▯

Theorem 11.10 is not true without the hypothesis that A be a Zariski ring. Consider, in fact, a regular ring A which is not factorial (Example 3) and let $\mathfrak{m}$ be a maximal ideal of A. Then the $\mathfrak{m}$-completion $(A, \mathfrak{m})\hat{}$ is a regular local ring, by 8.3 and 2.20, and hence factorial.

However, by combining Mori's Theorem and Nagata's theorem we may give the following criterion.

Theorem 11.11. *Let A be a ring, $\mathfrak{m}$ an ideal of A such that $1+\mathfrak{m}$ is generated by prime elements and suppose that $\bigcap \mathfrak{m}^n = 0$. Then, if the $\mathfrak{m}$-completion $\hat{A}$ of A is a factorial domain, A is also factorial.*

Proof. If $S = 1+\mathfrak{m}$, the Zariskification $S^{-1}A$ is factorial by 11.10; then A is factorial by 11.4 (notice that, with our hypothesis, $A \to \hat{A}$ is injective, hence A is an integral domain). ▯

Factoriality is not, in general, an ascending property from A to $\hat{A}$. In fact, even the property of being an integral domain may not be preserved by $\mathfrak{m}$-completions. Nevertheless, even if integrity is preserved, factoriality may not be (as we shall see later). We want now to give some conditions for the factoriality of $\hat{A}$ which are not in general related to the factoriality of A. We shall apply criterion 11.8 together with a preliminary lemma (Lemma 11.12) concerning relations between $\operatorname{Pic} A$, $\operatorname{Pic}(A/\mathfrak{m})$ and $\operatorname{Pic} \hat{A}$.

We notice that if $f: A \to B$ is a ring homomorphism, then we have a canonical homomorphism $f: \operatorname{Pic} A \to \operatorname{Pic} B$ (see [5], p. 145).

Lemma 11.12. *Let A be a ring, $\mathfrak{m}$ un ideal of A such that $\mathfrak{m} \subset \operatorname{rad} A$, $\hat{A}$ the $\mathfrak{m}$-completion of A. Then:*

a) *the homomorphism* $\varphi: \operatorname{Pic} A \to \operatorname{Pic}(A/\mathfrak{m})$ *is injective;*
b) *furthermore, if A is complete,* φ *is bijective;*
c) *the homomorphism* $\psi: \operatorname{Pic} A \to \operatorname{Pic} \hat{A}$ *is injective.*

Proof. See [1], Prop. 1.4, Prop. 1.7, Th. 1.18. □

Now we are able to give the criterion for the factoriality of $\hat{A}$ we have already announced. We need, for our purpose, only property b) of the lemma.

Theorem 11.13. *Let A be a ring,* $\mathfrak{m}$ *un ideal of A,* $\hat{A}$ *the* $\mathfrak{m}$*-completion of A. Then* $\hat{A}$ *is factorial if and only if the following properties are true:*
a) $\hat{A}$ *is locally factorial;*
b) $\operatorname{Spec}(A/\mathfrak{m})$ *is connected;*
c) $\operatorname{Pic}(A/\mathfrak{m}) = 0$.

Proof. We have $\operatorname{Pic}(A/\mathfrak{m}) \cong \operatorname{Pic}(\hat{A}/\mathfrak{m}\hat{A}) \cong \operatorname{Pic} \hat{A}$ by 11.12, b). The conclusion then follows by 11.8 and 10.8. □

Remark. Suppose that A is a regular ring; then we know that $\hat{A}$ is also regular (Theorem 8.3), hence condition a) of 11.13 is verified. Then, if $\mathfrak{m}$ is an ideal of A such that conditions b), c) of 11.13 are verified, the ring $\hat{A}$ is factorial. We have such a situation if A is a polynomial ring in n indeterminates $X_1, \ldots, X_n$ over a regular and factorial ring and $\mathfrak{m}$ is the ideal generated by the indeterminates; then $\hat{A}$ is the ring of formal power series over the basic ring. In this case $\hat{A}$ is factorial and we get the ascent of factoriality from A to $\hat{A}$ (see Theorem 11.14 below).

We notice that this ascent of factoriality from the polynomial ring $A[X_1, \ldots, X_n]$ to its $(X_1, \ldots, X_n)$-completion will be also an ascent of factoriality from A to $A[[X_1, \ldots, X_n]]$. As the problem of the ascent of factoriality from A to $A[[X_1, \ldots, X_n]]$ has an importance in itself, at least for historical reasons, the next section will be devoted to it.

C) Factorization in the ring of power series

First we give the mentioned criterion of ascent from a regular and factorial ring A. This result was given indipendently by P. Samuel ([35], Th. 2.1) and D. Buchsbaum ([12], Th. 3.2.).

Theorem 11.14. *Let A be a regular and factorial ring. Then* $A[[X]]$ *is also a factorial ring.*

Proof. We know, by 8.11, that $A[[X]]$ is a regular ring, hence locally factorial. As A is factorial, $\operatorname{Spec} A$ is connected and $\operatorname{Pic} A=0$; hence we may apply 11.13 to the pair $(A[X],(X))$ and get the result. □

In general, the factoriality of A does not imply the factoriality of $A[[X]]$. The first counterexample was given by Samuel in [35] as a consequence of the following result.

Theorem 11.15. *Let A be a domain, x, y, z three elements of A, and i, j, k three integers. We assume: y is prime, y and z are relatively prime,*

$$x^{i-1}\notin yA+zA, \quad x^i\in y^kA+z^jA, \quad ijk-ij-jk-ik\geqslant 0.$$

Then $A[[T]]$ *is not factorial (T being an indeterminate).*

Proof. See [35], Th. 4.1. □

In order to produce a factorial ring A such that $A[[T]]$ is not factorial, it is sufficient to find a factorial ring A, three elements x,y,z in A and three integers i,j,k satisfying the conditions of 11.15. Now, take a triple of integers i,j,k, two by two relatively prime, and suppose that the triple does not coincide with any permutation of the integers 2, 3, 5. Then it is easy to see that, if k is a field, then the ring $A=k[X,Y,Z]/(X^i+Y^j+Z^k)=k[x,y,z]$, the elements x,y,z and the integers i,j,k satisfy the hypothesis of 11.15; furthermore the ring A is factorial (see [8], p. 99, ex. 7) and we are done.

We notice that there is no difference in our conclusion if *we take for A the localization in the maximal ideal* (x,y,z) of the ring $k[x,y,z]$ above. Now, if T is an indeterminate, the localisation $A[T]_{(x,y,z,T)}$ of $A[T]$ in the maximal ideal (x,y,z,T) is a factorial ring whose completion $\hat{A}[[T]]$ is not factorial. Moreover, at least in the case $k=\mathbb{C}$, even the completion $\hat{A}$ of A is not a factorial ring: this is a consequence of a general result of Brieskorn (see [11], Satz 3.3 and Kor. 3.4) which shows that only with a permutation of the triple 2, 3, 5 do we get a factorial completion $\hat{A}=k[[x,y,z]]$.

Thus, we have seen that *there exists a non complete local noetherian factorial ring A, of dimension and depth both equal to* 2, *such that:*

α) *the ring* $A[[T]]$ *is not factorial;*

β) *the completion* $\hat{A}$ *of A is not factorial.*

Now a natural question arises. *Can* α) *hold with the further assumption that A is complete? The answer is still affirmative*, as is shown in an example given by P. Salmon, (see [32] and [33], n. 4, Esempio 1). However we notice the following general result due to G. Scheja.

Theorem 11.16. *If A is a complete local, noetherian factorial ring such that depth* $A \geqslant 3$, *then* $A[[T]]$ *is factorial.*

Proof. See [38], Satz 2. □

Further properties about factorization in power series rings can be found in [37].

D) Factorization in the ring of restricted power series. The Picard group of the ring $A[T]$

We recall that the ring of restricted power series $(A, \mathfrak{m})\{X_1, \ldots, X_n\}$ is the $\mathfrak{m}(X_1, \ldots, X_n)$-completion of the ring $B = A[X_1, \ldots, X_n]$. Hence, by 11.13, we see that factoriality of $(A, \mathfrak{m})\{X_1, \ldots, X_n\}$ is related to the vanishing of Pic $A[X_1, \ldots, X_n]/\mathfrak{m}(X_1, \ldots, X_n)$. Moreover, we can see that the above factoriality is also related to the vanishing of Pic $(A/\mathfrak{m})[X_1, \ldots, X_n]$, as is shown by the following criterion.

Theorem 11.17. *Let A be a regular factorial ring,* $\mathfrak{m}$ *an ideal of A such that* Pic $(A/\mathfrak{m})[X_1, \ldots, X_n] = 0$. *Then the ring* $(A, \mathfrak{m})\{X_1, \ldots, X_n\}$ *is factorial.*

Proof. See [1], Cor. 4.9. □

The above theorem is not true if we omit the assumption Pic $(A/\mathfrak{m})[X_1, \ldots, X_n] = 0$. For example, let k be a field, A the ring $k[X, Y]$, or its localisation $k[X, Y]_{(X,Y)}$ or the (X, Y)-completion $k[[X, Y]]$, and let $\mathfrak{m}$ be the principal ideal generated by the polynomial $X^i + Y^j$, where i, j are integers such that $ij - i - j > 0$. *Then the ring* $(A, \mathfrak{m})\{T\}$ *is non factorial* (see [30],

Remarque following Th. 9). Thus, by theorem 11.3, we have

$$(\gamma) \qquad \operatorname{Pic}\,(A/\mathfrak{m})[T] \neq 0.$$

The above considerations suggest the following remark: if A is a regular and factorial ring and $\mathfrak{m}$ an ideal of A, the vanishing of the Picard group of the ring $(A/\mathfrak{m})[X_1, \ldots, X_n]$ gives us the factoriality of the ring $(A, \mathfrak{m})\{X_1, \ldots, X_n\}$; on the other hand from the non-factoriality of the ring $(A, \mathfrak{m})\{X_1, \ldots, X_n\}$ we get the non-vanishing of the Picard group of the polynomial extension $(A/\mathfrak{m})[X_1, \ldots, X_n]$ of $A/\mathfrak{m}$. Finally, we could see that the vanishing of the Picard group of polynomial extensions is related to the notion of "weak normality" (see Prop. 3.3 in [14], which has been generalised recently by C. Traverso). Here, we restrict ourselves to mentioning the following result which is in some sense a generalisation of the above property (γ): *Let A be the local ring $k[X, Y]_{(X,Y)}$ (or $k[[X, Y]]$), $\mathfrak{m}$ the principal ideal of a simple plane curve with a singularity in the origin. Then we have* $\operatorname{Pic}\,(A/\mathfrak{m})[T] = 0$ *if and only if the origin is a node.* (See [34]).

The factoriality of restricted power series is also related to the factoriality of formal power series. As an example of this fact, we have the following result.

Theorem 11.18. *Let A be a complete local ring, $\mathfrak{m}$ its maximal ideal. Then, if $A[[T]]$ is factorial, the ring $(A, \mathfrak{m})\{T\}$ is also factorial.*

Proof. See [30], Th. 12. □

Similarly, non factoriality of restricted power series may be related with the analogous situation for formal power series. For example, the following result is also a consequence of Theorem 11.15: *Let A be the local ring at the origin of the surface $X^i + Y^j + Z^k$, where i, j, k are integers such that $ijk - ij - ik - jk \geqslant 0$; let $\mathfrak{m}$ be the maximal ideal of A. Then $(A, \mathfrak{m})\{T\}$ is not factorial.* (See [30], Coroll. to Th. 8).

§ 12
Fibers of a ring homomorphism and formal fibers of a ring

In this section we recall some results on Grothendieck's theory of fibers and formal fibers, which are very useful in the study of ascent of several local properties, such as, for example, R_n, S_n, reducedness and normality.

Let $\varphi: A \to B$ be a ring homomorphism, and let $f: \operatorname{spec} B \to \operatorname{spec} A$ be the continuous map induced by φ (f is defined by $f(\mathfrak{p}) = \varphi^{-1}(\mathfrak{p})$ for any $\mathfrak{p} \in \operatorname{spec} B$). If $\mathfrak{p} \in \operatorname{spec} A$ it is easy to see that there is a canonical bijection between $f^{-1}(\mathfrak{p})$ and $X = \operatorname{spec} S^{-1}(B/\mathfrak{p}B)$ where $S = A - \mathfrak{p}$. Thus X is called the *fiber* of f at $\mathfrak{p}$.

(By this technique one can give $f^{-1}(\mathfrak{p})$ a natural scheme structure, which has a great interest in algebraic geometry. We refer to [20] for a general treatment of this subject).

Notice that, if $k(\mathfrak{p})$ denotes the residue field of the local ring $A_\mathfrak{p}$ (which is isomorphic to the quotient field of the domain $A/\mathfrak{p}$), there is a canonical isomorphism $S^{-1}(B/\mathfrak{p}B) \cong B \otimes_A k(\mathfrak{p})$. Thus we shall say, by abuse of language, that the $k(\mathfrak{p})$-algebra $B \otimes_A k(\mathfrak{p})$ is the *fiber* of φ at $\mathfrak{p}$.

Example 12.1. Let A be a ring, $X_1 \ldots, X_n$ indeterminates over A, $\varphi: A \to A[X_1, \ldots, X_n]$ the canonical embedding. Let $\mathfrak{p} \in \operatorname{spec} A$. Then the fiber of φ at $\mathfrak{p}$ is canonically isomorphic to $k(\mathfrak{p})[X_1, \ldots, X_n]$, as one sees immediately.

If A is a local ring, the fibers of the canonical homomorphism $\varphi: A \to \hat{A}$ are called *formal fibers* of A. If A is any ring, and $\mathfrak{p}$ is a prime ideal of A, the *formal fibers* of A at $\mathfrak{p}$ are the formal fibers of the local ring $A_\mathfrak{p}$.

Example 12.2. Let A be a discrete valuation ring, $\mathfrak{m}$ its maximal ideal, k its residue field and K its quotient field. We have $k=k(\mathfrak{m})$ and $K=k(0)$. Thus the formal fiber of A at $\mathfrak{m}$ is $\hat{A}\otimes_A k\cong k$, and the formal fiber of A at (0) is $\hat{A}\otimes_A k$, which is canonically isomorphic to the quotient field $\hat{K}$ of $\hat{A}$, as one sees easily.

It is very useful to know the behaviour of the formal fibers of a ring A. In particular it is interesting to know whether they have "geometrical" properties, according to the following definition.

Definition 12.3. Le k be a field. A k-algebra A is said to be *geometrically regular* (or *normal, reduced* ...) if for any *finite* field extension k' of k the k'-algebra $A\otimes_k k'$ is regular (resp. normal, reduced etc.).

Since the fibers of a ring homomorphism are algebras over fields, the above definition can be applied to them.

Example 12.4. The k-algebra $k[X_1,\dots,X_n]=A$ is geometrically regular, since $A\otimes_k k'\cong k'[X_1,\dots,X_n]$ for any field extension k' of k, and is therefore regular. Thus, if A is a ring and $A[X_1,\dots,X_n]$ is a polynomial ring, the fibers of the canonical homomorphism $A\to A[X_1,\dots,X_n]$ are geometrically regular (see example 12.1).

Example 12.5. *If* $K\supset k$ is a field extension, K is geometrically regular if and only if K is separable over k (see e. g. [22], p. 148, 6.7.6). It follows then by example 12.2 that if A is a discrete valuation ring of characteristic 0, the formal fibers of A are geometrically regular.

Example 12.6. If A is a Dedekind domain of characteristic 0 (for example $\mathbb{Z}$), the formal fibers of A at any prime ideal are geometrically regular. The proof is similar to the preceding one.

The next theorem, due to Grothendieck, describes how geometrical properties of the formal fibers are preserved by the so called "essentially finitely generated" algebras (an A-Algebra B is called *essentially finitely generated* if $B=S^{-1}C$, where C is a finitely generated A-algebra, and S is a multiplicative set of C).

Theorem 12.7. *Let A be a noetherian ring. Then the following conditions are equivalent:*

(i) *the formal fibers of A at any maximal ideal are geometrically regular (resp. normal, reduced);*

(ii) *for any A-algebra B, essentially finitely generated, the formal fibers of B at any prime ideal are geometrically regular (resp. normal or reduced).*

Proof. The proof of this theorem is given in the greatest generality in [22] (p. 199, th. 7.4.4), for a large class of properties, which satisfy a certain number of conditions. It is not difficult to see that this is the case for geometric regularity, normality, reducedness (see e. g. [22], p. 194, 7.3.8). □

From the above theorem we can get many examples of rings whose formal fibers are geometrically regular, as the following corollary shows.

Corollary 12.8. *Let A be either a noetherian complete local ring (in particular a field) or a Dedekind domain of characteristic* 0, *and let B be an essentially finitely generated A-algebra. Then the formal fibers of A are geometrically regular.*

Proof. If A is a complete local ring, it is clear that its formal fibers are geometrically regular. On the other hand, if A is a Dedekind domain of characteristic zero, this fact follows from example 12.6. The conclusion follows then by Theorem 12.7. □

We end this section by recalling another result which relates the formal fibers of a noetherian ring A to the fibers of the canonical homomorphism $A \to \hat{A}$, where $\hat{A}$ is an $\mathfrak{m}$-completion of A.

Proposition 12.9. *Let A be a noetherian* $\mathfrak{m}$*-ring and* $\hat{A}$ *its completion. Suppose that the formal fibers of A at any maximal ideal containing* $\mathfrak{m}$ *are geometrically regular (resp. normal, reduced). Then the fibers of the canonical homomorphism* $A \to \hat{A}$ *are geometrically regular (etc...).*

Proof. This result, like Theorem 12.7, is also proved for a wider class of properties (see [22], p. 202, prop. 7.4.6), and is essentially a consequence of Proposition 6.2. □

§ 13
Properties S_n and R_n for $\mathfrak{m}$-completions

In this section all rings are assumed to be noetherian.

Definition 13.1. A ring A is said to be S_n $(n=0,1,\ldots)$ if $\operatorname{depth} A_\mathfrak{p} \geqslant \min\{n, \dim A_\mathfrak{p}\}$ for any $\mathfrak{p} \in \operatorname{spec} A$. This is equivalent to saying that $A_\mathfrak{p}$ is CM if $\dim A_\mathfrak{p} \leqslant n$ (see Th. 9.2), and $\operatorname{depth} A_\mathfrak{p} < n$ otherwise.

Examples. a) A ring is CM if and only if it is S_n for any n.

b) A ring A is S_1 if and only if its zero ideal is unmixed. It follows that an ideal $\mathfrak{a}$ of a ring A is unmixed if and only if $A/\mathfrak{a}$ is S_1.

c) Property S_2 has the following geometrical meaning: let V be an irreducible algebraic variety, $\mathcal{O}$ its structure sheaf, and $A=\Gamma(V,\mathcal{O})$ its coordinate ring. Then A is S_2 if and only if for any closed subvariety W of codimension >1, the restriction map $\Gamma(V,\mathcal{O}) \to \Gamma(V-W,\mathcal{O})$ is surjective (that is any rational function defined on $V-W$ can be extended to a rational function on V). This follows, for example, by [26], Prop. 2.

Now we recall some elementary results about the behaviour of property S_n with respect to flat and faithfully flat ring homorphisms.

Theorem 13.2. *Let A, B be two local rings, $\mathfrak{m}$ the maximal ideal of A and $\varphi: A \to B$ a flat local homomorphism. Then we have:*

(i) $\dim B = \dim A + \dim B/\mathfrak{m}B$.

(ii) $\operatorname{depth} B = \operatorname{depth} A + \operatorname{depth} B/\mathfrak{m}B$.

Proof: See [13] Theorems 5.1 and 5.2. □

Corollary 13.3. *Under the assumptions of Theorem* 13.2 *we have: B is CM if and only if A and $B/\mathfrak{m}B$ are CM.*

Proof. It follows easily from Theorem 13.2, and conditions (iv) of Theorem 9.2. □

Theorem 13.4. *Let $\varphi: A \to B$ be a flat ring homomorphism. Then we have for any n:*

(i) *If φ is faithfully flat and B is S_n, then A is S_n.*

(ii) *If A and the fibers of φ are S_n, then B is S_n.*

Proof. Let $\mathfrak{p} \in \operatorname{spec} A$. In order to prove (i) we must show that $\operatorname{depth} A_\mathfrak{p} \geqslant \min\{n, \dim A_\mathfrak{p}\}$. Since φ is faithfully flat, there is a $\mathfrak{q} \in \operatorname{spec} B$ such that $\varphi^{-1}(\mathfrak{q}) = \mathfrak{p}$. We may suppose that $\mathfrak{q}$ is minimal with respect to this property, so that $\dim B_\mathfrak{q}/\mathfrak{p}B_\mathfrak{q} = 0$. Let now $\varphi: A_\mathfrak{p} \to B_\mathfrak{q}$ be the flat local homomorphism induced by φ. Then by Theorem 13.3 we have: $\dim A_\mathfrak{p} = \dim B_\mathfrak{q}$ and $\operatorname{depth} A_\mathfrak{p} = \operatorname{depth} B_\mathfrak{q}$, and (i) follows immediately.

Let now $\mathfrak{q}$ be a prime ideal of B and $\mathfrak{p} = \varphi^{-1}(\mathfrak{q})$. Then one sees easily that $B_\mathfrak{q}/\mathfrak{p}B_\mathfrak{q}$ is a localization of the fiber of φ at $\mathfrak{p}$; the conclusion (ii) follows then from Theorem 13.2, applied to the flat local homomorphisms $A_\mathfrak{p} \to B_\mathfrak{q}$ induced by φ.

Condition (i) of the above theorem shows that property S_n descends by faithful flatness. Thus Proposition 6.3 and 6.5 can be applied to give corresponding statements on $\mathfrak{m}$-completions. It is also easy to see that by the mentioned propositions and the above Theorem we get the following theorem.

Theorem 13.5. *Let A be an $\mathfrak{m}$-ring, $\hat{A}$ its completion and n an integer. Suppose further that the following condition holds:*

a) *For any maximal ideal $\mathfrak{n}$ of A containing $\mathfrak{m}$, the formal fibers of $A_\mathfrak{n}$ are S_n.*

Then the following conditions are equivalent:

(i) *$\hat{A}$ is S_n;*

(ii) *$A_\mathfrak{n}$ is S_n for any maximal ideal $\mathfrak{n}$ containing $\mathfrak{m}$.*

If moreover $\mathfrak{m} \subset \operatorname{rad} A$, the above conditions are equivalent to:

(iii) *A is S_n.*

Now we want to show that condition a) of theorem 13.5 is verified for any n if A is a quotient of a CM ring. This will follow by our next proposition.

Proposition 13.6. *Let B be a CM local ring, $\mathfrak{b}$ an ideal of B and $A = B/\mathfrak{b}$. Then the formal fibers of A are CM.*

Proof. By Theorem 4.6 we have $\hat{A} \cong \hat{B} \otimes_B A \cong \hat{B}/\mathfrak{b}\hat{B}$, and hence the formal fibers of A are canonically isomorphic to the formal fibers of B at the prime ideals of B containing $\mathfrak{b}$. Let $\mathfrak{p}$ be such a prime ideal, and let $C \cong S^{-1}(\hat{B}/\mathfrak{p}\hat{B})$ be the formal fiber of B at $\mathfrak{p}$ ($S = B - \mathfrak{p}$). We must prove that $C_{\mathfrak{q}}$ is a CM local ring for any prime ideal $\mathfrak{q}$ of C (Th. 9.2), and we shall prove this fact by using Corollary 13.3. It follows from the definition of C that there is a $\mathfrak{q}' \in \operatorname{spec} \hat{B}$ such that $\mathfrak{q} = \mathfrak{q}'C$, and that the contraction of $\mathfrak{q}'$ to B is just $\mathfrak{p}$. Thus, since $\hat{B}$ is flat over B, we have a flat local homomorphism $B_{\mathfrak{p}} \to \hat{B}_{\mathfrak{q}}$, and since $C_{\mathfrak{q}} \cong B_{\mathfrak{q}'}/\mathfrak{p}\hat{B}_{\mathfrak{q}'}$, and $\hat{B}_{\mathfrak{q}'}$ is CM (Th. 9.3), the conclusion follows from Corollary 13.3. ∎

Corollary 13.7. *The conclusion of Theorem* 13.5 *is true for any* n *if the condition* a) *is replaced by:*

a′) *A is a quotient ring of a CM ring.*

An interesting consequence of the above Corollary is the following proposition.

Proposition 13.8. *Let A be a Zariski ring, $\hat{A}$ its completion, and $\mathfrak{a}$ an ideal of A. Suppose further that A is a CM ring. Then $\mathfrak{a}$ is unmixed if and only if $\hat{\mathfrak{a}}$ is unmixed.*

Proof. Since $(A/\mathfrak{a})\hat{} \cong \hat{A}/\hat{\mathfrak{a}}$ the conclusion follows from Corollary 13.7 applied to condition S_1 (see example b)). ∎

Now we turn our attention to property R_n. Te behaviour of this property is very similar to that of S_n, with the sole exception of Proposition 13.6. We recall first the definition and a few examples.

Definition 13.9. A ring A is said to be R_n ($n = 0, 1, \ldots$) if $A_{\mathfrak{p}}$ is a regular local ring whenever $\dim A_{\mathfrak{p}} \leqslant n$.

Examples. a) A ring is regular if and only if it is R_n for all n. Thus a regular ring is R_n and S_n for any n.

b) As we shall see in the following sections, a ring is reduced if and only if it is R_0 and S_1 and is normal if and only if it is R_1 and S_2.

c) It is known from algebraic geometry that the coordinate ring of an irreducible variety V is R_n if and only if the locus of singular points of V has codimension $> n$.

In order to study the ascent and descent of property R_n with respect to flat and faithfully flat ring homomorphisms, we need a result on regular local rings.

Proposition 13.10. *Let $\varphi: A \to B$ be a flat ring homomorphism, and let $\mathfrak{m}$ be the maximal ideal of A. Then:*

(i) *if B is regular, so is A.*

(ii) *If A and $B/\mathfrak{m}B$ are regular, B is regular.*

Proof. Since φ is faithfully flat (Prop. 4.8, iv), (i) follows immediately from Proposition 8.7. The other assertion follows easily from the equality $\dim B = \dim A + \dim B/\mathfrak{m}B$ (Th. 13.2) and the characterization of regular local rings given by Theorem 8.1, condition (iii). ▯

Theorem 13.11. *Let $\varphi: A \to B$ be a flat ring homomorphism and let n be an integer. Then:*

(i) *if B is R_n, and φ is faithfully flat, then A is R_n.*

(ii) *If A and all the fibers of φ are R_n, then B is R_n.*

Proof. It follows from Proposition 13.10, with the same argument as that used in the proof of Theorem 13.4. ▯

By the above result we can get, as for property S_n, the usual consequences about completions. In particular we shall need the following statement.

Theorem 13.12. *Theorem* 13.5. *remains true if we replace S_n by R_n.*

Proof. It follows from Theorem 13.11 with the same method as that used in the proof of Theorem 13.5. ▯

Corollary 13.13. *Let A be an essentially finitely generated algebra over a complete local ring or over a Dedekind domain of characteristic zero. Let $\mathfrak{m}$ be an ideal of A and $\hat{A}$ the $\mathfrak{m}$-completion of A. Then the following conditions are equivalent:*

(i) *$\hat{A}$ is R_n* (resp. S_n)

(ii) *A is R_n* (resp. S_n) *for any maximal ideal $\mathfrak{n}$ containing $\mathfrak{m}$.*

If moreover $\mathfrak{m} \subset \operatorname{rad} A$, the above conditions are equivalent to:

(iii) *A is R_n* (resp. S_n).

Proof. The formal fibers of A are geometrically regular (and hence R_n and S_n for any n) by Corollary 12.8. The conclusion follows then by Theorems 13.5 and 13.12. □

Remark. The part of Corollary 13.13 relative to property S_n could also be deduced by Corollary 13.7. In fact a ring A satisfying the hypothesis of Corollary 13.13 is a quotient of a regular local ring. This is clear for essentially finitely generated algebras over Dedekind domains, while for complete local ring it follows by the famous Cohen Structure Theorem, (see, e.g., [13], th. 6.3).

§ 14
Analytic reducedness

In this section we shall give some sufficient conditions for the reducedness of $\mathfrak{m}$-adic completions which are related to the radical of the completion of an ideal. *In this section all rings are assumed to be noetherian.*

We say that a local ring A is *analytically reduced* if $\hat{A}$ is reduced, i.e. if it has no nilpotent elements. In this case A itself is reduced, since it is isomorphic to a subring of $\hat{A}$ by Krull's intersection theorem (Th. 1.5).

A first sufficient condition for reducedness of $\mathfrak{m}$-completions is the following proposition, which is an immediate consequence of Proposition 6.5.

Proposition 14.1. *Let A be an $\mathfrak{m}$-ring and $\hat{A}$ its completion. Suppose that $A_{\mathfrak{n}}$ is analytically reduced for any maximal ideal $\mathfrak{n}$ containing $\mathfrak{m}$. Then $\hat{A}$ is reduced.*

In order to apply the above proposition we have to see whether a local ring is analytically reduced. This is not always true, as examples given by Nagata show (see [29], Appendix). This can be done by using the formal fibers and the following characterization of reducedness in terms of properties R_0 and S_1 (for an easy proof, see e.g. [13], prop. 4.5, p. 73).

Proposition 14.2. *A ring A is reduced if and only if it is R_0 and S_1.*

By the above proposition we have results for reducedness similar to Theorems 13.5 and 13.12. Then one can get easily:

Proposition 14.3. *Let A be an $\mathfrak{m}$-adic ring and $\hat{A}$ its completion. Suppose that $A_{\mathfrak{n}}$ and its formal fibers are reduced for any maximal ideal $\mathfrak{n}$ containing $\mathfrak{m}$. Then $\hat{A}$ is reduced.*

We can apply the above proposition to the radical of an ideal, as the following Theorem shows.

Theorem 14.4. *Let A be an $\mathfrak{m}$-ring, $\mathfrak{a}$ an ideal of $A, \hat{A}$ and $\hat{\mathfrak{a}}$ the $\mathfrak{m}$-completions of A and $\mathfrak{a}$ respectively. Suppose that the formal fibers of $A_\mathfrak{n}$ are reduced for any maximal ideal $\mathfrak{n}$ containing $\mathfrak{m}$. Then we have: $\sqrt{\hat{\mathfrak{a}}} = (\sqrt{\mathfrak{a}})\hat{}$.*

Proof. Since $\hat{\mathfrak{a}} = \mathfrak{a}\hat{A}$ for any ideal of A (Th. 4.5) it is clear that $(\sqrt{\mathfrak{a}})\hat{} \subset \sqrt{\hat{\mathfrak{a}}}$; thus in order to prove the opposite inclusion it is sufficient to show that $\hat{A}/(\sqrt{\mathfrak{a}})\hat{}$ is reduced, or, equivalently that the $\mathfrak{m}$-completion of $A' = A/\sqrt{\mathfrak{a}}$ is reduced. This follows immediately from Proposition 14.3, since it is not difficult to see that for any maximal ideal $\mathfrak{n}'$ of A' containing $\mathfrak{m}A'$, the formal fibers of $A'_{\mathfrak{n}'}$ are reduced. □

Corollary 14.5. *Under the assumptions of Theorem* 14.4, *we have:*

(i) *If* $\mathfrak{a} = \sqrt{\mathfrak{a}}$, *then* $\hat{\mathfrak{a}} = \sqrt{\hat{\mathfrak{a}}}$. *In particular if* $\mathfrak{p}$ *is prime, then* $\hat{\mathfrak{p}} = \sqrt{\hat{\mathfrak{p}}}$.

(ii) $(A_{\mathrm{red}})\hat{} \cong \hat{A}_{\mathrm{red}}$ (where by B_{red} we mean the ring B modulo its nilradical).

A special case of rings which satisfy the hypothesis on formal fibers of Theorem 14.4 are the "Pseudo-geometric rings" defined by Nagata (and called "Universellement Japonaises" by Grothendieck). We shall call then simply Nagata rings.

Definition 14.6. A ring A is said to be a *Nagata ring* if for any $\mathfrak{p} \in \operatorname{spec} A$ and any finite field extension K of $k(\mathfrak{p})$, the integral closure of $A/\mathfrak{p}$ in K is a finitely generated $(A/\mathfrak{p})$-module (cfr. [29], p. 131 and [21], p. 213, 23.11)

Nagata local rings can be characterised in terms of their formal fibers by means of the following theorem, which is essentially a restatement, due to Grothendieck, of important results due to Zariski and Nagata, about analytic reducedness of essentially finitely generated local rings over a field and over a Dedekind domain of characteristic zero respectively.

Theorem 14.7. *Let A be a semilocal ring. Then the following conditions are equivalent:*

(i) *A is a Nagata ring;*

(ii) *The formal fibers of A at any maximal ideal are geometrically reduced.*

Proof. See [22], (p. 209, 7.6.4 and p. 212 Th. 7.7.2). □

It is not difficult to see that if A is a Nagata ring, all its rings of fractions $S^{-1}A$ are Nagata rings. In particular the formal fibers of Nagata rings are geometrically reduced. Thus we have immediately the following Corollary.

Corollary 14.8. *If A is a Nagata ring, the conclusion of theorem* 14.4 *and Corollary* 14.5 *are true.*

Remark 1. It can be shown that the essentially finitely generated algebras over a complete local domain and over Dedekind domains of characteristic zero are Nagata rings. To be more precise they are *excellent rings*, which are always Nagata rings. We refer to [22], p. 182 for more details on formal fibers, Nagata rings and excellent rings.

Remark 2. It is not known whether an $\mathfrak{m}$-adic completion of a Nagata ring (or an excellent ring) is still a Nagata (or an excellent) ring. More precisely it is not known what happens to the formal fibers when we pass from a ring A to an $\mathfrak{m}$-adic completion of A (see e.g. [22], p. 203). However, it can be shown that the *henselization* of a Nagata (resp. excellent) ring A with respect to an ideal $\mathfrak{m}$ is a Nagata (resp. excellent) ring. This is known for local rings (see [23], p. 143, cor. 18.7.3, and p. 144, Cor. 16.7.6), and has been recently proved (and partially published in [19]) by S. Greco in the general case.

§ 15
Normality of $\mathfrak{m}$-completions

Let B a ring and A a subring of B. An element $x \in B$ is said to be *integral* over A if there are $a_0, \ldots, a_{n-1} \in A$ such that $a_0 + \cdots + a_{n-1} x^{n-1} + x^n = 0$ $(n > 0)$. The ring A is said to be *integrally closed in B* if every element of B which is integral over A is an element of A. Finally a domain A is said to be *integrally closed* if A is integrally closed in its quotient field.

Definition 15.1. A ring A is said to be *normal* if and only if $A_\mathfrak{p}$ is an integrally closed domain for any $\mathfrak{p} \in \operatorname{spec} A$.

Examples. a) A domain is normal if and only if it is integrally closed (see e.g. [3], p. 63, prop. 5.13).

b) A noetherian ring is normal if and only if it is a direct product of integrally closed domains. This follows easily from the definition and Lemma 10.3.

c) A factorial ring is integrally closed (see e.g. [3], p. 63). It follows that any locally factorial ring is normal. In particular any regular ring is normal (Th. 11.9).

Now we recall Serre's normality criterion for noetherian rings.

Theorem 15.2. *A noetherian ring is normal if and only if it is R_1 and S_2.*

Proof. See, e.g. [22], p. 108, Th. 5.8.6. □

By Serre's criterion and Theorems 13.5 and 13.11 we get immediately the following result.

Corollary 15.4. *Let $\varphi: A \to B$ be a flat homomorphism of noetherian rings. Then:*

(i) *If B is normal and φ is faithfully flat, A is normal.*

(ii) *If A and the fibers of φ are normal, then B is normal.*

Now we return to $\mathfrak{m}$-adic completions. We say that a local ring A is *analytically normal* if its completion $\hat{A}$ is normal. If A is *noetherian*, then A itself must be normal by Theorem 4.9 and Corollary 15.4. (i).

By the usual techniques we get the following theorems.

Theorem 15.5. *Let A be a noetherian $\mathfrak{m}$-ring and $\hat{A}$ its completion. Then:*

(i) *If $A_\mathfrak{n}$ is analytically normal for any maximal ideal $\mathfrak{n}$ containing $\mathfrak{m}$, then $\hat{A}$ is normal.*

(ii) *If moreover* $\operatorname{spec}(A/\mathfrak{m})$ *is connected, then $\hat{A}$ is a normal domain.*

Proof. The first part follows from Proposition 6.5, by means of the usual techniques. The second follows easily by Proposition 10.8 and Example a). □

Theorem 15.6. *Let A be a noetherian $\mathfrak{m}$-ring, and $\hat{A}$ its completion. Suppose further that the formal fibers of A are normal for any maximal ideal $\mathfrak{n}$ containing $\mathfrak{m}$. Then the following conditions are equivalent:*

(i) *$\hat{A}$ is normal;*

(ii) *$A_\mathfrak{n}$ is normal for any maximal ideal $\mathfrak{n}$ containing $\mathfrak{m}$;*

If moreover $\mathfrak{m} \subset \operatorname{rad} A$, *then the above conditions are equivalent to:*

(iii) *A is normal.*

Proof. It follows by 13.5, 13.12 and 15.2. □

Remark. Since a regular ring is normal, the above theorem can be applied to all rings described in Corollary 12.8, since they have geometrically regular formal fibers.

Now we want to apply the above results to the rings of restricted power series. We need a well known lemma.

Lemma 15.7. *Let A be a ring. Then the polynomial ring $A[X_1, \ldots, X_n]$ is normal if and only if A is normal.*

Proof. It is an easy consequence of [7], p. 20, Cor. 3. □

Proposition 15.8. *Let A be a noetherian $\mathfrak{m}$-ring and $B=(A,\mathfrak{m})\{X_1,\dots,X_n\}$ the ring of restricted power series. Then we have:*

(i) *If B is normal, A is normal.*

(ii) *If A is normal and its formal fibers at any maximal ideal are geometrically normal, then B is normal.*

Proof. Since B is faithfully flat over A, (Cor. 4.10), (i) follows from Corollary 15.4. Conversely if A is normal, $A[X_1,\dots,X_n]$ is normal by Lemma 15.7. Moreover since the formal fibers of A are geometrically normal, so are the fibers of $A[X_1,\dots,X_n]$ (Th. 12.7); and since B is the $\mathfrak{m}(X_1,\dots,X_n)$-completion of $A[X_1,\dots,X_n]$, the conclusion follows from Theorem 15.6.

Remark. The conclusion of Theorem 15.6 is not true, in general, without the assumption on the formal fibers (for counterexamples see, e.g. [29], p. 208—209). It should be remarked, however, that: if *A is a normal noetherian ring, then the ring $A[[X_1,\dots,X_n]]$ is normal* (as follows by [7], p. 20, n. 4). Thus if $C=A[X_1,\dots,X_n]$ is normal, the $(X_1,\dots,X_n)$-adic completion of C is normal, *without any assumption on the formal fibers of C*. It is therefore reasonable to believe that assertion (i) of the above Proposition must be true without any assumption on the formal fibers.

We conclude by observing that if, instead of $\mathfrak{m}$-completion we consider $\mathfrak{m}$-henselization, normality does ascend in general, also for non noetherian rings, (see e.g. [25], prop. 8 for the local case, and [18], Cor. 7.6 for the general case).

References

1. Arezzo, D., Greco, S.: Sul gruppo delle classi di ideali. Ann. Scuola Norm. Sup. Pisa, Cl. di Sc. XXI, Fasc. IV, 459–483 (1967).
2. — Robbiano L.: Sul completamento di un anello rispetto ad un ideale di tipo finito. Rend. Sem. Mat. Univ. Padova, 1970 (to appear).
3. Atiyah, M. F., MacDonald I. G.: Introduction to Commutative Algebra. Reading: Addison Wesley, 1969.
4. Bass, H.: On the ubiquity of Gorenstein rings. Math. Z. **82**, 8—28, (1963).
5. Bourbaki, N.: Algèbre Commutative. Ch. I, II. Paris: Hermann 1961.
6. — Algèbre Commutative. Ch. III, IV. Paris: Hermann, 1961.
7. — Algèbre Commutative. Ch. V, VI. Paris: Hermann, 1964.
8. — Algèbre Commutative. Ch. VII. Paris: Hermann, 1965.
9. — Topologie Générale. Ch. I, II, 4ème èd. Paris: Hermann 1965.
10. — Topologie Générale. Ch. III, IV. Paris: Hermann 1960.
11. Brieskorn, E.: Rationale Singularitäten komplexer Flächen. Invent. Math. **4**, 336—358 (1968).
12. Buchsbaum, D.: Some remarks on factorization in power series rings. J. Math. Mech. **10**, 749—753 (1961).
13. Dieudonné, J.: Topics in local Algebra. Notes by Mario Borelli. Notre Dame Math. Lect. n. 10. Notre Dame, Indiana: University of Notre Dame Press, 1967.
14. Endo, S.: Projective modules over polynomial rings. J. Math. Soc. Japan **15**, n. 3, 339—352 (1963).
15. Greco, S., Salmon, P.: Anelli di Macaulay. Pubblicazioni dell'Istituto Matematico dell'Università di Genova, 1965.
16. — Sugli ideali frazionari invertibili. Rend. Sem. Mat. Univ. Padova **36**, 315—333 (1966).
17. — Anelli di Gorenstein. Seminario Ist. Mat. Univ. Genova 1969.
18. — Henselization of a ring with respect to an ideal. Trans. Amer. Math. Soc. **144**, 43—65 (1969).
19. — Sugli omorfismi piatti e non ramificati. Le Matematiche, Catania, **XXIV**, fasc. 2, 392—415 (1969).
20. Grothendieck, A., Dieudonné, J.: Eléments de Géométrie Algébrique. I. I. H. E. S. Publ. Math. n° 4, 1960.

21. — — Eléments de Géométrie Algébrique. IV. (Première partie), Publ. Math. n° 20, 1964.
22. — — Eléments de Géométrie Algébrique. IV. (Seconde partie), I. H. E. S. Publ. Math. n° 24, 1965.
23. — — Eléments de Géométrie Algébrique. IV. I. H. E. S. (Quatrième partie), Publ. Math. n° 32, 1967.
24. Kaplanski, I.: Commutative Rings. Boston; Allen and Bacon, 1970.
25. Lafon, J. P.: Anneaux henseliens. Bull. Soc. Math. France **91**, 77—107 (1963).
26. Margaglio, C.: Alcune proprietà delle R-coppie in un domino di integrità. Rend. Sem. Mat. Univ. Padova **XXXIX**, 389—399 (1967).
27. Millevoi, T.: Sulle estensioni di un anello di Gorenstein. Rend. Sem. Mat. Univ. Padova **XLI**, 319—325 (1968).
28. Murthy, P. M.: A note on factorial rings. Arch. Math. **15**, 418—420 (1964).
29. Nagata, M.: Local Rings. New York: Interscience 1962.
30. Salmon, P.: Sur les séries formelles restreintes. Bull. Soc. Math. France **92**, 385—410 (1964).
31. — Serie convergenti su un corpo non archimedeo con applicazione ai fasci analitici. Ann. Mat. Pura Appl., Serie IV **LXV**, 113—125 (1964).
32. — Su un problema posto da P. Samuel. Rend. Acc. Lincei, Cl. di Sc., Serie VIII **XL**, fasc. 5, 801—803 (1966).
33. — Sulla fattorialità delle algebre graduate e degli anelli locali. Rend. Sem. Mat. Univ. Padova **XLI**, 119—137 (1968).
34. — Singolarità e gruppo di Picard. Istituto Nazionale di Alta Matematica, Symposia Mathematica **II**, 341—345 (1968).
35. Samuel, P.: On unique factorization domains. Illinois J. Math. **5**, 1—17 (1961).
36. — Anneaux factoriels, Rédaction de A. Micali, Soc. de Mat. de Sao Paulo, 1963.
37. — Lectures on unique factorization domains. Tata Institute of fundamental research, Bombay, 1964.
38. Scheja, G.: Einige Beispiele faktorieller lokaler Ringe. Math. Ann. **172**, 124—134 (1967).
39. Serre, J. P.: Algèbre locale. Multiplicitées. Lecture notes in math. n. 11. Berlin-Heidelberg-New York: Springer 1965.
40. Tate, J.: Rigid analytic spaces. Private notes reproduced with (out) his permission by I. H. E. S., Paris 1962.
41. Zariski, O., Samuel, P.: Commutative Algebra. Vol. II Princeton: Van Nostrand 1960.

Subject Index

Ergebnisse der Mathematik und ihrer Grenzgebiete

1. Bachmann: Transfinite Zahlen. DM 42,—; US $ 11.60
2. Miranda: Partial Differential Equations of Elliptic Type. DM 58,—; US $ 16.00
4. Samuel: Méthodes d'Algèbre Abstraite en Géométrie Algébrique. DM 29,—; US $ 8.00
5. Dieudonné: La Géométrie des Groupes Classiques. DM 42,—; US $ 11.60
6. Roth: Algebraic Threefolds with Special Regard to Problems of Rationality. DM 22,—; US $ 6.10
7. Ostmann: Additive Zahlentheorie. 1. Teil: Allgemeine Untersuchungen. DM 42,—; US $ 11.60
8. Wittich: Neuere Untersuchungen über eindeutige analytische Funktionen. DM 31,—; US $ 8.60
11. Ostmann: Additive Zahlentheorie. 2. Teil: Spezielle Zahlenmengen. DM 28,—; US $ 7.70
13. Segre: Some Properties of Differentiable Varieties and Transformations. DM 46,—; US $ 12.70
14. Coxeter/Moser: Generators and Relations for Discrete Groups. DM 35,—; US $ 9.70
15. Zeller/Beekmann: Theorie der Limitierungsverfahren. DM 64,—; US $ 17.60
16. Cesari: Asymptotic Behavior and Stability Problems in Ordinary Differential Equations. 3rd edition in preparation
17. Severi: Il teorema di Riemann-Roch per curve-superficie e varietà questioni collegate. DM 26,—; US $ 7.20
18. Jenkins: Univalent Functions and Conformal Mapping. DM 37,—; US $ 10.20
19. Boas/Buck: Polynomial Expansions of Analytic Functions. DM 18,—; US $ 5.00
20. Bruck: A Survey of Binary Systems. DM 46,— ;US $ 12.70
21. Day: Normed Linear Spaces. DM 17,80; US $ 4.90
23. Bergmann: Integral Operators in the Theory of Linear Partial Differential Equations. DM 40,—; US $ 11.00
25. Sikorski: Boolean Algebras. DM 42,—; US $ 11.60
26. Künzi: Quasikonforme Abbildungen. DM 43,—; US $ 11.90
27. Schatten: Norm Ideals of Completely Continuous Operators. DM 26,—; US $ 7.20
28. Noshiro: Cluster Sets. DM 40,—; US $ 11.00
29. Jacobs: Neuere Methoden und Ergebnisse der Ergodentheorie. DM 55,—; US $ 15.20
30. Beckenbach/Bellman: Inequalities. DM 38,—; US $ 10.50
31. Wolfowitz: Coding Theorems of Information Theory. DM 30,—; US $ 8.30
32. Constantinescu/Cornea: Ideale Ränder Riemannscher Flächen. DM 75,—; US $ 20.70
33. Conner/Floyd: Differentiable Periodic Maps. DM 29,—; US $ 8.00
34. Mumford: Geometric Invariant Theory. DM 24,—; US $ 6.60
35. Gabriel/Zisman: Calculus of Fractions and Homotopy Theory. DM 42,—; US $ 11.60
36. Putnam: Commutation Properties of Hilbert Space Operators and Related Topics. DM 31,—; US $ 8.60
37. Neumann: Varieties of Groups. DM 51,—; US $ 14.10
38. Boas: Integrability Theorems for Trigonometric Transforms. DM 20,—; US $ 5.50
39. Sz.-Nagy: Spektraldarstellung linearer Transformationen des Hilbertschen Raumes. DM 20,—; US $ 5.50
40. Seligman: Modular Lie Algebras. DM 43,—; US $ 11.90
41. Deuring: Algebren. DM 26,—; US $ 7.20
42. Schütte: Vollständige Systeme modaler und intuitionistischer Logik. DM 26,—; US $ 7.20
43. Smullyan: First-Order Logic. DM 36,—; US $ 9.90
44. Dembowski: Finite Geometries. DM 68,—; US $ 17.00

45. Linnik: Ergodic Properties of Algebraic Fields. DM 44,—; US $ 12.10
46. Krull: Idealtheorie. DM 28,—; US $ 7.70
47. Nachbin: Topology on Spaces of Holomorphic Mappings. DM 18,—; US $ 5.00
48. A. Ionescu Tulcea/C. Ionescu Tulcea: Topics in the Theory of Lifting. DM 36,—; US $ 9.90
49. Hayes/Pauc: Derivation and Martingales. DM 48.—; US $ 13.20
50. Kahane: Séries de Fourier Absolument Convergentes. DM 44,—; US $ 12.10
51. Behnke/Thullen: Theorie der Funktionen mehrerer komplexer Veränderlichen. DM 48,—; US $ 13.20
52. Wilf: Finite Sections of Some Classical Inequalities. DM 28,—; US $ 7.70
53. Ramis: Sous-ensembles analytiques d'une variété banachique complexe. DM 36,—; US $ 9.90
54. Busemann: Recent Synthetic Differential Geometry. DM 32,—; US $ 8.80
55. Walter: Differential and Integral Inequalities. DM 74,—; US $ 20.40
56. Monna: Analyse non-archimédienne. DM 38,—; US $ 10.50
57. Alfsen: Compact Convex Sets and Boundary Integrals. In preparation
58. Greco/Salmon: Topics in m-Adic Topologies. DM 24,—; US $ 6.60
59. Medrano, de: Involutions on Manifolds. DM 36,—; US $ 9.90
60. Sakai: C^*-Algebras and W^*-Algebras. In preparation

Ergebnisse der Mathematik und ihrer Grenzgebiete

---

1. Bachmann: Transfinite Zahlen. DM 42,—; US $ 11.60
2. Miranda: Partial Differential Equations of Elliptic Type. DM 58,—; US $ 16.00
4. Samuel: Méthodes d'Algèbre Abstraite en Géométrie Algébrique. DM 29,—; US $ 8.00
5. Dieudonné: La Géométrie des Groupes Classiques. DM 42,—; US $ 11.60
6. Roth: Algebraic Threefolds with Special Regard to Problems of Rationality. DM 22,—; US $ 6.10
7. Ostmann: Additive Zahlentheorie. 1. Teil: Allgemeine Untersuchungen. DM 42,—; US $ 11.60
8. Wittich: Neuere Untersuchungen über eindeutige analytische Funktionen. DM 31,—; US $ 8.60
11. Ostmann: Additive Zahlentheorie. 2. Teil: Spezielle Zahlenmengen. DM 28,—; US $ 7.70
13. Segre: Some Properties of Differentiable Varieties and Transformations. DM 46,—; US $ 12.70
14. Coxeter/Moser: Generators and Relations for Discrete Groups. DM 35,—; US $ 9.70
15. Zeller/Beekmann: Theorie der Limitierungsverfahren. DM 64,—; US $ 17.60
16. Cesari: Asymptotic Behavior and Stability Problems in Ordinary Differential Equations. 3rd edition in preparation
17. Severi: Il teorema di Riemann-Roch per curve-superficie e varietà questioni collegate. DM 26,—; US $ 7.20
18. Jenkins: Univalent Functions and Conformal Mapping. DM 37,—; US $ 10.20
19. Boas/Buck: Polynomial Expansions of Analytic Functions. DM 18,—; US $ 5.00
20. Bruck: A Survey of Binary Systems. DM 46,—; US $ 12.70
21. Day: Normed Linear Spaces. DM 17,80; US $ 4.90
23. Bergmann: Integral Operators in the Theory of Linear Partial Differential Equations. DM 40,—; US $ 11.00
25. Sikorski: Boolean Algebras. DM 42,—; US $ 11.60
26. Künzi: Quasikonforme Abbildungen. DM 43,—; US $ 11.90
27. Schatten: Norm Ideals of Completely Continuous Operators. DM 26,—; US $ 7.20
28. Noshiro: Cluster Sets. DM 40,—; US $ 11.00
29. Jacobs: Neuere Methoden und Ergebnisse der Ergodentheorie. DM 55,—; US $ 15.20
30. Beckenbach/Bellman: Inequalities. DM 38,—; US $ 10.50
31. Wolfowitz: Coding Theorems of Information Theory. DM 30,—; US $ 8.30
32. Constantinescu/Cornea: Ideale Ränder Riemannscher Flächen. DM 75,—; US $ 20.70
33. Conner/Floyd: Differentiable Periodic Maps. DM 29,—; US $ 8.00
34. Mumford: Geometric Invariant Theory. DM 24,—; US $ 6.60
35. Gabriel/Zisman: Calculus of Fractions and Homotopy Theory. DM 42,—; US $ 11.60
36. Putnam: Commutation Properties of Hilbert Space Operators and Related Topics. DM 31,—; US $ 8.60
37. Neumann: Varieties of Groups. DM 51,—; US $ 14.10
38. Boas: Integrability Theorems for Trigonometric Transforms. DM 20,—; US $ 5.50
39. Sz.-Nagy: Spektraldarstellung linearer Transformationen des Hilbertschen Raumes. DM 20,—; US $ 5.50
40. Seligman: Modular Lie Algebras. DM 43,—; US $ 11.90
41. Deuring: Algebren. DM 26,—; US $ 7.20
42. Schütte: Vollständige Systeme modaler und intuitionistischer Logik. DM 26,—; US $ 7.20
43. Smullyan: First-Order Logic. DM 36,—; US $ 9.90
44. Dembowski: Finite Geometries. DM 68,—; US $ 17.00

45. Linnik: Ergodic Properties of Algebraic Fields. DM 44,—; US $ 12.10
46. Krull: Idealtheorie. DM 28,—; US $ 7.70
47. Nachbin: Topology on Spaces of Holomorphic Mappings. DM 18,—; US $ 5.00
48. A. Ionescu Tulcea/C. Ionescu Tulcea: Topics in the Theory of Lifting. DM 36,—; US $ 9.90
49. Hayes/Pauc: Derivation and Martingales. DM 48.—; US $ 13.20
50. Kahane: Séries de Fourier Absolument Convergentes. DM 44,—; US $ 12.10
51. Behnke/Thullen: Theorie der Funktionen mehrerer komplexer Veränderlichen. DM 48,—; US $ 13.20
52. Wilf: Finite Sections of Some Classical Inequalities. DM 28,—; US $ 7.70
53. Ramis: Sous-ensembles analytiques d'une variété banachique complexe. DM 36,—; US $ 9.90
54. Busemann: Recent Synthetic Differential Geometry. DM 32,—; US $ 8.80
55. Walter: Differential and Integral Inequalities. DM 74,—; US $ 20.40
56. Monna: Analyse non-archimédienne. DM 38,—; US $ 10.50
57. Alfsen: Compact Convex Sets and Boundary Integrals. In preparation
58. Greco/Salmon: Topics in $\mathfrak{m}$-Adic Topologies. DM 24,—; US $ 6.60
59. Medrano, de: Involutions on Manifolds. DM 36,—; US $ 9.90
60. Sakai: C^*-Algebras and W^*-Algebras. In preparation